New Studies in Biology

Biotechnology

Second Edition

John E. Smith

B.Sc., M.Sc., Ph.D., D.Sc., F.I. Biol., F.R.S.E.
Professor of Applied Microbiology
University of Strathclyde

Edward Arnold
A division of Hodder & Stoughton
LONDON   BALTIMORE   MELBOURNE   AUCKLAND

© 1988 John E. Smith

First published in Great Britain 1981

British Library Cataloguing Publication Data

Smith, John E. (John Edward), *1932–*
  Biotechnology. — 2nd ed.
  1. Biotechnology
  I. Title   II. Series
  660′.6

  ISBN 0-7131-2960-3

Typeset in 10/11pt Plantin by Colset Private Limited, Singapore
Printed and bound in Great Britain for Edward Arnold, the educational, academic and medical publishing division of Hodder and Stoughton Limited, 41 Bedford Square, London WC1B 3DQ by Whitstable Litho Ltd, Whitstable, Kent.

General Preface to the Series

Recent advances in biology have made it increasingly difficult for both students and teachers to keep abreast of all the new developments in so wide-ranging a subject. The New Studies in Biology, originating from an initiative of the Institute of Biology, are published to facilitate resolution of this problem. Each text provides a synthesis of a field and gives the reader an authoritative over-view of the subject without unnecessary detail.

The Studies series originated 20 years ago but its vigour has been maintained by the regular production of new editions and the introduction of additional titles and new themes become clearly identified. It is appropriate for the New Studies in their refined format to appear at a time when the public at large has become conscious of the beneficial applications of knowledge from the whole spectrum from molecular to environmental biology. The new series is set to provide as great a boon to the new generation of students as the original series did to their fathers.

1986   Institute of Biology
                                                        20 Queensberry Place
                                                        London SW7 2DZ

Preface

Biotechnology is in reality a series of enabling technologies and involves the practical application of biological organisms, or their subcellular components, to manufacturing and service industries and to environmental management. Biotechnology is a subject of great antiquity having its origins in ancient microbial processes such as brewing, wine making and fermented milk products, such as cheeses and yoghurts. However, new developments in enzyme technology, fermentation processes, monoclonal antibodies, and in genetic engineering, have introduced new and exciting dimensions to the subject. It is becoming increasingly important to develop in young people an awareness of the existence and future role of biotechnology in modern society. This new and much expanded edition of *Biotechnology* is aimed to give an integrated overview of biotechnology and for some students to point the way forward for exciting and satisfying careers.

I am deeply indebted to Miss Liz Clements for typing the manuscript.

I dedicate this edition to my grandaughters Christy, Emma, Lucy and Kim.

John E. Smith
1988

Contents

1

An Introduction to Biotechnology

1.1 What is biotechnology?

There is little doubt that modern biology is the most diversified of all the Natural sciences. It exhibits a bewildering array of subdisciplines, microbiology, plant and animal anatomy, biochemistry, immunology, cell biology, plant and animal physiology, morphogenesis, systematics, ecology, palaeobotany, genetics and many others. The increasing diversity of modern biology derived primarily from the largely post war introduction into biology of other scientific disciplines, such as physics, chemistry and mathematics, which have made possible the description of life processes at the cellular and molecular level. In the last two decades well over twenty Nobel prizes have been awarded for discoveries in these fields of study.

This newly acquired biological knowledge has already made vastly important contributions to the health and welfare of man. And yet, what has gone before may well pale into insignificance if all the hopes of *biotechnology* can be realized.

Biotechnology has been defined in many forms (Table 1.1) but in essence implies the use of microbial, animal or plant cells or enzymes to synthesize, break down or transform materials. This requires the integration of biochemistry, biology, microbiology, chemical engineering and process engineering, together with other disciplines, in a way that optimizes the exploitation of their potential (Fig. 1.1). Biotechnology is not itself a product or range of products like microelectronics: rather it should be regarded as a range of enabling technologies which will find significant application in several industrial sectors. As will be seen in later sections, it is a technology in search of new applications and the main benefits lie in the future. New biotechnological processes will, in most instances, function at low temperature, will consume little energy and will rely mainly on inexpensive substrates for biosynthesis.

However, it should be recognized that biotechnology is not new but represents a developing and expanding series of technologies with roots established (in many cases) thousands of years ago. Biotechnology includes many traditional processes such as brewing, baking, winemaking, cheesemaking, the production of oriental foods such as soy sauce and tempeh, and sewage treatment where the use of microorganisms has been developed somewhat

Table 1.1 · Some selected definitions of biotechnology.

'The application of biological organisms, systems or processes to manufacturing and service industries.'

'The integrated use of biochemistry, microbiology and engineering sciences in order to achieve technological (industrial) application of the capabilities of microorganisms, cultured tissue cells and parts thereof.'

'A technology using biological phenomena for copying and manufacturing various kinds of useful substances.'

'The application of scientific and engineering principles to the processing of materials by biological agents to provide goods and services.'

'The science of the production processes based on the action of microorganisms and their active components and of production processes involving the use of cells and tissues from higher organisms. Medical technology, agriculture and traditional crop breeding are not generally regarded as biotechnology.'

'Biotechnology is really no more than a name given to a set of techniques and processes.'

'Biotechnology is the use of living organisms and their components in agriculture, food and other industrial processes.'

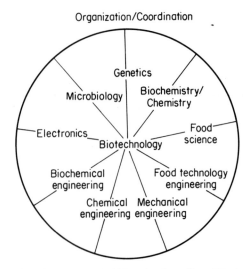

Fig. 1.1   The interdisciplinary nature of biotechnology (from Higgins *et al,*, 1985).

empirically over many years (Table 1.2). It is only recently that these processes have been subjected to rigorous scientific scrutiny and analysis; even so it will surely take some time for modern scientifically based practices fully to replace traditional empiricism.

There is also a considerable danger that biotechnology will be viewed as a coherent, unified body of scientific and engineering knowledge and thinking, to

Table 1.2 Historical development of biotechnology.

1 *Biotechnological production of foods and beverages*
Sumerians and Babylonians were drinking beer by 6000 BC; Egyptians were baking leavened bread by 4000 BC; wine was known in the Near East by the time of the book of Genesis. Microorganisms first seen in 17th century by Anton van Leeuwenhoek who developed the simple microscope; fermentative ability of microorganisms demonstrated between 1857–1876 by Pasteur – *the father of biotechnology*; cheese production has ancient origins, also mushroom cultivation.

2 *Biotechnological processes initially developed under nonsterile conditions*
Ethanol, acetic acid, butanol and acetone produced by end of 19th century by open microbial fermentation processes; waste-water treatment and municipal composting of solid wastes, the largest fermentation capacity practised throughout the world.

3 *Introduction of sterility to biotechnological processes*
Introduction in the 1940s of complicated engineering techniques to the mass cultivation of microorganisms to exclude contaminating microorganisms. Examples include antibiotics, amino acids, organic acids, enzymes, steroids, polysaccharides and vaccines.

4 *Applied genetics and recombinant DNA technology*
Traditional strain improvement of important industrial organisms has long been practised; recombinant DNA techniques together with protoplast fusion allow new programming of the biological properties of organisms.

be applied in a coherent and logical manner. This is not so; the range of biological, chemical and engineering disciplines that are involved have varying degrees of application to the industrial scene.

Traditional or 'old' biotechnology has established an expanding market. 'New' or modern aspects of biotechnology, founded in recent advances in molecular biology, genetic engineering and fermentation process technology, are not yet finding wide industrial application. A wide range of knowledge and expertise is ready to be put to productive use; but the rate at which it will be applied will depend less on scientific or technical considerations and more on such factors as adequate investment by the relevant industries, improved systems of biological patenting, marketing skills and the economics of the new methods in relation to technologies currently employed.

The industrial activities to be affected include human and animal food production, provision of chemical feedstocks to replace petrochemical sources, alternative energy sources, waste recycling, pollution control and agriculture. The new techniques will also revolutionize many aspects of medicine, veterinary science and pharmaceutics.

Biotechnological industries will be based largely on renewable and recyclable materials and so can be adapted to the needs of a society in which energy is ever-increasingly expensive and scarce. In many ways, biotechnology is an embryonic technology and will require much skilful control of its development, but the potentials are vast and diverse, and it undoubtedly will play an increasingly important part in many future industrial processes.

1.2 Biotechnology — an interdisciplinary pursuit

Biotechnology is *a priori* an interdisciplinary pursuit. In recent decades a characteristic feature of the development of science and technology has been the increasing resort to multidisciplinary strategies for the solution of various problems. This has led to the emergence of new interdisciplinary areas of study, with the eventual crystallization of new disciplines with identifiable, characteristic concepts and methodologies.

Chemical engineering and biochemistry are two well-recognized examples of disciplines that have done much to clarify our understanding of chemical processes and the chemical basis of biological systems.

The term *multidisciplinary* describes a quantitative extension of approaches to problems that commonly occur within a given area. It involves the marshalling of concepts and methodologies from a number of separate disciplines and applying them to a specific problem in another area. In contrast, *interdisciplinary* application occurs when the blending of ideas that occur during multidisciplinary cooperation leads to the crystallization of a new disciplinary area with its own concepts and methodologies. In practice, multidisciplinary enterprises are almost invariably mission-orientated. However, when true interdisciplinary synthesis occurs the new area will open up a novel spectrum of investigations. Biotechnology has arisen through the interaction between various parts of biology and engineering.

A biotechnologist employs techniques derived from chemistry, microbiology, biochemistry, chemical engineering and computer science (Fig. 1.1). The main objectives are the innovation, development and optimal operation of processes in which biochemical catalysis has a fundamental and irreplaceable role. Biotechnologists must also aim to achieve a close working cooperation with experts from other related fields such as medicine, nutrition, the pharmaceutical and chemical industries, environmental protection and waste process technology.

The application of biotechnology will increasingly rest upon the ability of each of the contributing disciplines to understand the technical language of the others and – above all – to understand the potential as well as the limitations of the other areas.

A key factor in the distinction between biology and biotechnology is the scale of operation. The biologist usually works in the range between nanograms and milligrams. The biotechnologist working on the production of vaccines may be satisfied with milligram yields, but in most other projects aims at kilograms or tonnes. Thus, one of the main aspects of biotechnology consists of scaling-up biological processes.

Many present-day biotechnological processes have their origins in ancient and traditional fermentations such as the brewing of beer and the manufacture of bread, cheese, yoghurt, wine and vinegar. However, it was the discovery of antibiotics in 1929 and their subsequent large-scale production in the 1940s that created the greatest advances in fermentation technology. Since then we have witnessed a phenomenal development in this technology, not only in the

Table 1.3 World markets for biological products, 1981.

Product	Sales (£ million)
Alcoholic beverages	23 000
Cheese	14 000
Antibiotics	4 500
Penicillins	500
Tetracyclines	500
Cephalosporins	450
Diagnostic tests	2 000
Immunoassay	400
Monoclonal	5
Seeds	1 500
High-fructose syrups	800
Amino acids	750
Baker's yeast	540
Steroids	500
Vitamins	
All	330
C	200
B_{12}	14
Citric acid	210
Enzymes	200
Vaccines	150
Human serum albumin	125
Insulin	100
Urokinase	50
Human factor VIII protein	40
Human growth hormone	35
Microbial pesticides	12

(From P. Dunnill (1984) SERC Biotechnology Directorate Newsletter, 1(1)).

production of antibiotics but in many other useful, simple or complex chemical products, for example organic acids, polysaccharides, enzymes, vaccines, hormones, etc. (Table 1.3). Inherent in the development of fermentation processes is the close relationship between the biochemist, the microbiologist and the chemical engineer. Thus, biotechnology is not a sudden discovery but rather a coming of age of a technology that was initiated several decades ago. Looking to the future, the *Economist* when reporting on this new technology stated that it may launch 'an industry as characteristic of the twenty-first century as those based on physics and chemistry have been of the twentieth century'.

If it is accepted that biotechnology has its roots in distant history and has large successful industrial outlets, why then has there been such public awareness of this subject in recent years? The main reason must derive from the rapid advances in molecular biology – in particular recombinant DNA technology – which are giving humans dominance over nature. By these new techniques (discussed in Chapter 4) it is possible to manipulate directly the heritable material (DNA) of cells between like and unlike cells, creating new combinations of characters and abilities not previously present on this planet. The potential of this series of techniques, first developed in academic laboratories, is being rapidly exploited in industry. The industrial benefits are immense but the inherent dangers of tampering with nature must always be appreciated and respected.

The developments of biotechnology are proceeding at a speed similar to that of the microelectronics industry in the mid-1970s. Although the analogy is tempting, any expectations that biotechnology will develop commercially at the same rate should be tempered with considerable caution. While the potential of 'new' biotechnology cannot be doubted, meaningful commercial realization and benefits are not expected until well into the 1990s. New biotechnology will have a considerable impact across a wide range of chemical substances. In each case the economics of competing means of production will influence the development of a biotechnological route. Biotechnology will undoubtedly have great benefits in the long term.

The growth in awareness of modern biotechnology parallels the serious worldwide changes in the economic climate arising from the escalation of oil prices since 1973. There is a growing realization that fossil fuels (although at present in a production glut period) and other non renewable resources will one day be in limited supply. This will result in the requirement of cheaper and more secure energy sources and chemical feedstocks, which biotechnology could perhaps fulfil. Countries with climatic conditions suitable for rapid biomass production could well have major economic advantages over less climatically suitable parts of the world. In particular, the tropics must hold high potential in this respect.

Another contributory factor to the growing interest in biotechnology has been the current world recession, in particular the depression of the chemical and engineering sections, in part due to increased energy costs. Biotechnology has been considered as one important means of restimulating the economy, whether on a local, regional, national or even global basis, using new biotechnological methods and new raw materials. In part, the industrial boom of the 1950s and 1960s was due to cheap oil; while the information technology advances in the 1970s and 1980s resulted from developments in microelectronics. It is quite feasible that the 1980s and 1990s will be the era of biotechnology. There is undoubtedly a worldwide increase in molecular biological research, the formation of new biotechnological companies, large investments by nations, companies and individuals and the rapid expansion of data bases, information sources and, above all, extensive optimistic media coverage.

It is perhaps unfortunate that there has been excessive concentration on the new implications of biotechnology and less interest paid to the very large biotechnological industrial bases that already function throughout the world and contribute considerably to most nations' gross national profits. Indeed, many of the innovations in biotechnology will not appear *a priori* as new products but rather as improvements to organisms and processes in long-established biotechnological industries, e.g. brewing and antibiotics production.

New applications are likely to be seen earliest in the area of health care and medicine, followed by agriculture and food technology. Applications in chemical production, fuel and energy production, pollution control and resource recovery will possibly take longer to develop and will depend on changes in the relative economics of currently employed technologies. Biotechnology-based industries will not be labour-intensive; and although they will create valuable new employment, the need will be more for brains than muscle.

Biotechnology is high technology *par excellence*. The most exciting and potentially profitable facets of new biotechnology in the 1990s will involve research and development at the very frontiers of current knowledge and techniques.

In the late 1970s biotechnologists were putting forward vague promises about the wonders of their subject, while the realizing technologies were still being developed and still required immense levels of research and product development fundings. Biotechnologists now make predictions with more confidence since many of the apparently insurmountable problems have been overcome more easily than had been predicted, and many transitions from laboratory experiments to large-scale industrial processes have been achieved. Truly, new biotechnology has come of age.

1.3 Biotechnology — a three-component central core

Many biotechnological processes may be considered as having a three-component central core, in which one part is concerned with obtaining the best biological catalyst for a specific function or process, the second part creates (by construction and technical operation) the best possible environment for the catalyst to perform, and the third part (downstream processing) is concerned with the separation and purification of an essential product or products from a fermentation process.

In the majority of examples so far developed, the most effective, stable and convenient form for the catalyst for a biotechnological process is a whole organism, and it is for this reason that so much of biotechnology revolves around microbial processes. This does not exclude the use of higher organisms; in particular, plant and animal cell culture will play an increasingly important role in biotechnology (Chapter 9).

Microorganisms can be viewed both as primary fixers of photosynthetic energy and as systems for bringing about chemical changes in almost all types of natural and synthetic organic molecules. Collectively, they have an immense gene pool which offers almost unlimited synthetic and degradative potential. Furthermore, microorganisms possess extremely rapid growth rates far in excess of any of the higher organisms such as plants and animals. Thus immense quantities can be produced under the right environmental conditions in short time periods.

The methodologies that are in general use enable the selection of improved microorganisms from the natural environmental pool, the modification of microorganisms by mutation and, more recently, the mobilization of a spectacular array of new techniques, deriving from molecular biology, which may eventually permit the construction of microorganisms and plants with totally novel biochemical potentials (Chapter 4). These new techniques have arisen from the pure scientific efforts in molecular biology over the last two decades.

These carefully selected and manipulated organisms must be maintained in substantially unchanged form, and this involves another spectrum of techniques for preserving organisms, for retaining essential features during industrial processes and, above all, for retaining vigour and viability. In many examples the catalyst is used in a separated and purified form, i.e. as enzymes; a huge amount of information has been built up on the large-scale production, isolation and purification of individual enzymes and on their stabilization by artificial means (Chapter 5).

The second part of this central core of biotechnology encompasses all aspects of the containment system or bioreactor within which the catalysts must function (Chapter 3). Here the specialist knowledge of the chemical or process engineer provides the design and instrumentation for maintaining and controlling the physicochemical environment – temperature, aeration, pH, etc. – thus allowing the optimum expression of the catalyst (Chapter 3).

Having achieved the successful formation of a biotechnological product within a bioreactor, in most cases it will be necessary to separate the product from the aqueous environment. This can be a technically difficult and expensive procedure, and is the least understood area of biotechnology. Down-

Table 1.4   The main areas of application of biotechnology.

1   *Fermentation technology*
Historically, the most important area of biotechnology (brewing, antibiotics, etc.); extensive development in progress with new products envisaged – polysaccharides, medically important drugs, solvents, protein-enhanced foods. Novel fermenter designs to optimize productivity.

2   *Enzyme technology*
Used for the catalysis of extremely specific chemical reactions; immobilization of enzymes; to create specific molecular converters (bioreactors). Products formed include L-amino acids, high-fructose syrup, semisynthetic penicillins, starch and cellulose hydrolysis, etc. Enzyme probes for analysis.

3   *Waste technology*
Long historical importance, but more emphasis now on coupling these processes with the conservation and recycling of resources; foods and fertilizers, biological fuels.

4   *Environmental technology*
Great scope exists for the application of biotechnological concepts for solving many environmental problems – pollution control, removing toxic wastes; recovery of metals from mining wastes and low-grade ores.

5   *Renewable resources technology*
The use of renewable energy sources, in particular, lignocellulose, to generate chemical raw materials and energy-ethanol, methane and hydrogen. Total utilization of plant and animal material.

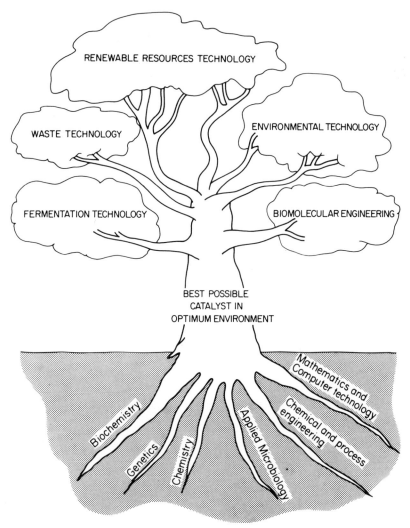

Fig. 1.2 Biotechnology tree.

stream processing is primarily concerned with initial separation of the bioreactor broth or medium into a liquid phase and a solids phase, and subsequent
concentration and purification of the product. Processing will usually involve
more than one stage. Downstream processing costs (as approximate proportions of selling prices) of fermentation products vary considerably; e.g. with
yeast biomass, penicillin G and certain enzymes, processing costs as percen-
tages of selling price are 20%, 20 to 30% and 60 to 70% respectively.

Successful involvement in a biotechnological problem therefore draws
heavily upon more than one of the input disciplines. The main areas of

application of biotechnology are shown in Table 1.4, while Fig. 1.2 shows how the many disciplines input into the biotechnological processes.

Biotechnology will continue to create exciting new opportunities for commercial development and profits in a wide range of industrial sectors, including health care and medicine, agriculture and forestry, fine and bulk chemicals production, food technology, fuel and energy production, pollution control and resource recovery. Biotechnology offers the hope of solving many of the problems our world faces.

In the following chapters some of the most important areas of biotechnology are considered with a view to achieving a broad overall understanding of the existing achievements and future aims of this new area of technology. However, it must be appreciated that biotechnological development will not only depend on scientific and technological advances, but will also be subject to considerable political and economic forces.

2

Substrates for Biotechnology

2.1 The nature of biomass

Plant biomass and (to a lesser extent), animal biomass represent utilizable sources of carbon for biotechnological processes. Historically there are well-known examples based on these sources, such as the production of alcohol from grain and cheese from milk. Primary photosynthetic productivity (growth of plants using solar energy) of the earth has been estimated to be 155×10^9 tonnes of material per year on a dry weight basis.

The highest proportion of land-based biomass (44%) is produced as forest (see Table 2.1). It is surprising to note that while agricultural crops account for only 6% of the primary photosynthetic productivity, from this amount is derived a major portion of food for humans and animals as well as many essential structural materials, textiles and paper products. Many traditional agricultural products may be further exploited with the increasing awareness of biotechnology. In particular, new technological approaches will undoubtedly be able to utilize the large volume of waste material that presently finds little use with conventional food processing.

Biomass agriculture and forestry may hold great economic potential for many national economies, particularly in tropical and subtropical regions (see Fig. 2.1). Indeed, the application of biotechnological processes in developing areas where plant growth excels could well bring about a change in the balance of economic power.

Table 2.1 Breakdown of world primary productivity.

	Net productivity (% of total)
Forests and woodlands	44.3
Grassland	9.7
Cultivated land	5.9
Desert and semidesert	1.5
Freshwater	3.2
Oceans	35.4

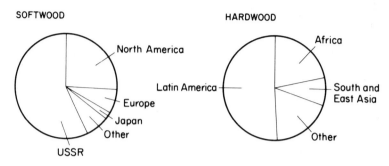

Fig. 2.1 The distribution of world forestry resources.

2.2　Natural raw materials

Natural raw materials originate from agriculture and forestry. These are mainly carbohydrates of varying chemical complexity and include sugar, starch, cellulose, hemicellulose and lignin. The wide range of byproducts obtained from raw materials and of use in biotechnological processes is shown in Table 2.2.

Sugar-bearing raw materials such as sugar beet, sugar cane and sugar millet are the most suitable and available to serve as feedstocks for biotechnological processing. As traditional uses of sugar are replaced by more efficient alternatives, the sugar surplus on the commodity market will give further incentive to develop new uses. Many tropical economies would collapse if the market for sugar were removed. Already cane sugar serves as the substrate for the Brazilian

Table 2.2　A range of byproducts that could be used as substrates in biotechnology.

Agriculture	Forestry	Industry
Straw	Wood waste hydrolysate	Molasses
Bagasse	Sulphite pulp liquor	Distillery wastes
Maize cobs	Bark, sawdust	Whey
Coffee, cocoa and coconut hulls	Paper and cellulose	Industrial waste water from food industries (olive,
Fruit peels and leaves	Fibres	palm-oil, potato, date,
Tea wastes		citrus, cassava)
Oilseed cakes		Wash waters (dairy,
Cotton wastes		canning, confectionery,
Bran		bakery, soft drinks,
Pulp (tomato, coffee, banana, pineapple, citrus, olive)		sizing, maltings, corn steep)
Animal wastes		Fishery effluent and wastes
		Meat byproducts
		Municipal garbage
		Sewage
		Abattoir wastes

'gasohol' programme (Chapter 7) and many other nations are rapidly seeing the immense potential of these new technologies.

Starch-bearing agricultural products include the various types of grain such as maize, rice and wheat, together with potatoes and other root crops such as sweet potato and cassava. A slight disadvantage of starch is that it must usually be degraded to monosaccharides or oligosaccharides by digestion or hydrolysis before fermentation. However, many biotechnological processes using starch are being developed, including fuel production.

There can be little doubt that cellulose, both from agriculture and forestry sources, must contribute a major source of feedstock for biotechnological processes such as fuels and chemicals. However, cellulose is a very complex chemical and invariably occurs in nature in close association with lignin. The ability of lignocellulose complexes to withstand the biodegradative forces of nature is witnessed by the longevity of trees, which are mainly composed of lignocellulose.

Lignocellulose is the most abundant and renewable natural resource available to man throughout the world. However, massive technological difficulties must be overcome before economic use may be made of this plentiful compound. At present, expensive energy-demanding pretreatment processes are required to open up this complex structure to wide microbial degradation. Pure cellulose can be degraded by chemical or enzymatic hydrolysis to soluble sugars which can be fermented to form ethanol, butanol, acetone, single cell protein (SCP), methane and many other products. Exciting advances are being made in the USA, Sweden and Britain and it is only a matter of time before these difficulties are overcome. It has been realistically calculated that approximately 3.3×10^{14} kg of CO_2 per year are fixed on the surface of the earth, and that approximately 6% of this, i.e. 22 billion tonnes per year, will be cellulose. On a worldwide basis, land plants produce 24 tonnes of cellulose per person per year. Time will surely show lignocellulose to be the most useful carbon source for biotechnological developments.

2.3 Availability of byproducts

A primary objective of biotechnology is to improve the management and utilization of the vast volumes of waste organic materials to be found throughout the world. The biotechnological utilization of these wastes will eliminate a source of pollution, in particular water pollution, and convert these wastes into useful byproducts.

Not all processes will involve biosystems. In particular, the processes of reverse osmosis and ultrafiltration are finding increasing uses. Reverse osmosis is a method of concentrating liquid solutions in which a porous membrane allows water to pass through but not the salts dissolved in it. Ultrafiltration is a method of separating the high and low molecular weight compounds in a liquid, by allowing the liquid and low molecular weight compounds to pass through while holding back the high molecular weight compounds and suspended solids. Some current applications of these technologies include concentration of dilute factory effluents; concentration of dilute food products; sterilization of water; purification of brackish water; and separation of edible solids from dilute effluents.

Waste materials are frequently important for economic and environmental reasons. For example, many byproducts of the food industry are of low economic value and are often discharged into waterways, creating serious environmental pollution problems. An attractive feature of carbohydrate waste as a raw material is that, if its low cost can be coupled with low handling costs, an economic process may be obtained. Furthermore, the worldwide trend towards stricter effluent control measures, or parallel increases in effluent disposal charges, can lead to the concept of waste as a 'negative cost' raw material. However, the composition or dilution of the waste may be such that transport costs to a production centre may be prohibitive. On these occasions biotechnology may only serve to reduce a pollution hazard.

Each waste material must be assessed for its suitability for biotechnological processing. Only when a waste is available in large quantities and preferably over a prolonged period of time can a suitable method of utilization be considered (Table 2.3).

Two widely occurring wastes that already find considerable fermentation uses are molasses and whey. Molasses is a byproduct of the sugar industry and has a sugar content of approximately 50%. Molasses is widely used as a fermentation feedstock for the production of antibiotics, organic acids and commercial yeasts for baking, and is directly used in animal feeding. Whey, obtained during the production of cheese, could also become a major fermentation feedstock.

More complex wastes such as straw and bagasse are widely available and their use increased as improved processes for lignocellulose breakdown become available (Table 2.4). Wood wastes include low-grade wood, bark and sawdust, as well as waste liquors such as sulphite waste liquor from pulp production which already finds considerable biotechnological processing in Europe and Soviet bloc countries (Chapter 6).

The largest proportion of total volume of waste matter is from animal rearing (faeces, urine), then agricultural wastes, wastes from food industries and finally

Table 2.3 Biotechnological strategies for utilization of suitable organic waste materials.

1 Upgrade the food waste quality to make it suitable for human consumption.

2 Feed the food waste directly or after processing to poultry, pigs, fish or other single-stomach animals that can utilize it directly.

3 Feed the food waste to cattle or other ruminants if unsuitable for single-stomach animals because of high fibre content, toxins or other reasons.

4 Production of biogas (methane) and other fermentation products if waste is unsuitable for feeding without expensive pretreatments.

5 Selective other purposes such as direct use as fuel, building materials, chemical extraction, etc.

Table 2.4 Pretreatments required before substrates are suitable for fermentation.

Substrate	Pretreatment
Sugary materials Sugar cane, beet, molasses, fruit juices, whey	Minimal requirements dilution and sterilization
Starchy materials Cereals, rice, vegetables, process liquid wastes	Some measure of hydrolysis by acid or enzymes. Initial separation of nonstarch components may be required.
Lignocellulosic materials Corn cobs, oat hulls, straw, bagasse, wood wastes, sulphite liquor, paper wastes	Normally requires complex pretreatment involving reduction in particle size followed by various chemical or enzymic hydrolyses. Energy-intensive and costly.

domestic wastes. The disposal of many waste materials, particularly animal wastes, is no problem in traditional agriculture and is particularly well exemplified in China where recycling by composting has long been practised. However, where intensive animal rearing is undertaken, serious pollution problems do arise.

2.4 Chemical and petrochemical feedstocks

With the development of commercial processes for the production of single cell protein (SCP) and other organic products, a number of chemical and petrochemical feedstocks have become particularly important for fermentation processes, since these materials have the advantage of being available in large quantities and in the same quality in most parts of the world. Thus, natural gas or methane and gas oil have been preferred as raw material because of their easy processing and universal availability. Main commercial interest has been concerned with *n*-paraffins, methanol and ethanol. Their involvement in various aspects of biotechnology, particularly in SCP production, are considered later. Table 2.5 summarizes the many technical considerations involved in the utilization of waste materials.

Future biotechnological processes will increasingly make use of organic materials that are renewable in nature or occur as low-value wastes that may cause environmental pollution. Some processes may also more economically utilize specific fractions of fossil fuels as feedstocks for biotechnological processes.

Table 2.5 Technical considerations for the utilization of waste materials.

Biological availability
  low (cellulosics)
  moderate (starch, lactose)
  high (molasses, pulping sugars)
Concentration
  solid (milling residues, garbage)
  concentrated (molasses)
  weak (lactose, pulping sugars)
  very dilute (process and plant wash liquors)
Quality
  clean (molasses, lactose)
  moderate (straw)
  dirty (garbage, feedlot waste)
Location
  collected (large installation, small centres)
  collected specialized (olive, palm oil, date, rubber, fruit/vegetable)
  dispersed (straw, forestry)
Seasonality
  prolonged (palm oil, lactose)
  very short (vegetable cannery waste)
Alternative uses
  some (straw)
  none (garbage)
  negative (costly effluents)
Local technology potential
  high (USA)
  middle (Brazil)
  low (Malaysia)

2.5 Raw materials and the future of biotechnology

The future development of large-scale biotechnological processes is inseparable from the supply and cost of raw materials. During the early and middle parts of this century the availability of cheap oil led to an explosive development of the petrochemical industry; many products formerly derived from the fermentation ability of microorganisms were superseded by the cheaper and more efficient chemical methods. However, escalating oil prices in the 1970s created profound reappraisals of these processes; and as the price of crude oil approaches that of some major cereal products (Fig. 2.2) there is a reawakening of interest in many fermentation processes for the production of ethanol and related products.

The most important criteria determining the selection of a raw material for a biotechnological process will include price, availability, composition, form and oxidation state of the carbon source. Table 2.6 gives the mid-1984 prices of existing and potential raw materials of biotechnological interest. At present the

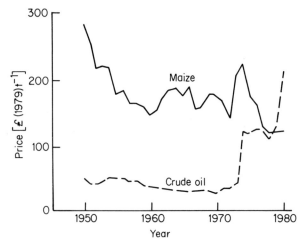

Fig. 2.2 Prices of maize and crude oil, 1950–1980 (from King, 1982).

Table 2.6 Prices of available raw materials for biotechnological processes.

Substrate	Mid-1984 US price ($ per t)	Carbon content (g mol C per mol substrate)	Carbon content relative to glucose (%)	Corrected price relative to glucose ($ per t)[a]
Corn starch	70–100[b]	0.44	100	64–91
Glucose	290[c]	0.4	100	290
Sucrose – raw	140[d]	0.42	105	133
Sucrose – refined	660[e]	0.42	105	140
Molasses	79	0.2[f]	50	140
Acetic acid	550	0.4	100	550
Methanol	150	0.375	94	160
Ethanol	560	0.52	130	430
Methane	n.a.	0.75	188	—
Corn oil (crude)	330	0.8	200	165
Palm oil	600	0.8	200	300
n – alkanes n – hexadecane)	n.a.	0.87	218	—

[a] Assumes equivalent conversion efficiencies can be obtained.
[b] Approximate guessed price in a wet-milling operation.
[c] Glucose syrups on a dry weight basis.
[d] Daily spot price.
[e] US price fixed by government tariffs.
[f] On the basis of molasses being 48% by weight fermentable sugars.
n.a. not available
(from Hacking, 1986).

most widely used and commercially valuable are corn starch, methanol, molasses and raw sugar.

There is little doubt that cereal crops – particularly maize, rice and wheat – will be the main short-term and medium-term raw materials for biotechnological processes. Furthermore, it is believed that this can be achieved without seriously disturbing human and animal food supplies. Throughout the world there is an uneven distribution of cereal production capacity and demand. Overproduction of cereals occurs mostly where extensive biotechnological processes are in practice. Areas of poor cereal production will undoubtedly benefit from the developments in agricultural biotechnology now at an early but highly promising stage.

Although much attention has been given to the use of wastes in biotechnology there are many major obstacles to be overcome. For instance, availability of agricultural wastes is seasonal and geographic availability problematic; they are also often dilute and may contain toxic wastes.

However, their build-up in the environment can present serious pollution problems and therefore their utilization in biotechnological processes, albeit at little economic gain, can have overall community value.

Although in the long term biotechnology must seek to utilize the components of cellulose and lignocellulose as fuels or feedstocks, the technological difficulties are considerable. The chemical complexities of these molecules are notorious and it is proving more difficult than expected to break them down economically to usable primary molecules.

3

Fermentation Technology

3.1 Introduction

Fermentation technology can involve complete living cells (microbes, animal and plant cells) or components of cells (enzymes) and is directed to effect specific chemical or physical changes on organic substances. It is not enough just to achieve the necessary chemical or physical changes; the method must also be competitive with other means of production such as chemical technology. In most cases these biologically driven processes will be used industrially because they are the only practical way in which the desired product(s) can be achieved.

The advantages of producing organic products by biological as opposed to purely chemical methods are many.

(1) Complex organic molecules such as proteins and antibiotics cannot practically be made by chemical methods.
(2) Bioconversions give higher yields.
(3) Biological systems operate at lower temperatures, near neutral pH, etc.
(4) There is greater specificity of catalytic reaction.
(5) Biological processes can achieve exclusive production of an isomeric compound.

There are also some distinct disadvantages in comparison with chemical methods.

(1) Biological systems can be easily contaminated with foreign, unwanted microorganisms.
(2) The desired product will be present in a complex product mixture requiring separation.
(3) It is necessary to provide, handle and dispose of large volumes of water.
(4) Bioprocesses are usually extremely slow when compared with conventional chemical processes.

For each biotechnological process a suitable containment system must be designed and the resultant biological processes monitored and controlled. For most practical purposes the system of containment is the fermenter or bioreactor, which provides the physical environment in which the desired biocatalysts can interact with the environment and material supply. Bioreactors range from simple stirred or unstirred tanks to complex integrated systems involving varying levels of computer input.

Bioreactors occur in two distinct types. The first type are nonaseptic systems where it is not absolutely essential to operate with strictly pure cultures of microorganisms, for example, brewing, baker's yeast production, and effluent disposal systems. In the second type aseptic operation is essential for the production of such compounds as antibiotics, amino acids, polysaccharides and SCP. In this category of bioreactor all other microorganisms must be excluded; this involves considerable challenges to the engineering construction and operation.

The physical form of many of the most widely used bioreactors has not altered much over the past thirty years; however, more recently, novel forms of bioreactor have been developed to suit the needs of specific bioprocesses, and such innovations will have increasingly active roles in biotechnology.

Within any bioreactor the primary aim is to optimize the growth of the organism or of a substance produced by the organism. To achieve this objective, optimum culture conditions must be developed and will involve the following requirements: an energy source; other essential nutrients to satisfy the needs of the organism; a lack of inhibiting compounds in the medium; a reliable inoculum; and the most advantageous physicochemical conditions.

Commercial fermentation processes are largely similar irrespective of what organism is selected, what medium is used and what product is formed. The primary aim of commercial processes is to produce large numbers of cells with uniform characteristics by growing them under defined, controlled conditions. Indeed the same bioreactors with only minor modifications can be used to produce an enzyme, an antibiotic, an organic acid or single cell protein. Some of the many products of biotechnology produced in bioreactors are listed in Table 3.1.

3.2 Media design for fermentation processes

Water is the dominant component in almost all current bioprocesses. In order to exploit the full potential of water-based biotechnological processes it is essential to optimize the composition of the growth media to be used at all stages of the process, i.e. inoculum preparation, seed cultures, and final production stages.

The specific needs of the organism can be easily determined at the laboratory level and scaled up to production level. When the process is on the scale of the large bioreactors, it is important that the materials used are readily available, reproducible, stable and easy to handle and store. Costs of materials are a major concern, since the successful development of any biotechnological process will, to a large extent, depend on its cost relative to other methods of preparing the same product. Furthermore, since world commodity prices fluctuate

Table 3.1 Fermentation products according to industrial sectors.

Sector	Activities
Chemicals	
Organic (bulk)	Ethanol, acetone, butanol
	Organic acids (citric, itaconic)
Organic (fine)	Enzymes
	Perfumeries
	Polymers (mainly polysaccharides)
Inorganic	Metal beneficiation, bioaccumulation and leaching (CU, U)
Pharmaceuticals	Diagnostic agents (enzymes, monoclonal antibodies)
	Enzyme inhibitors
	Steroids
	Vaccines
Energy	Ethanol (gasohol)
	Methane (biogas)
Food	Dairy products (cheeses, yoghurts, fish and meat products)
	Beverages (alcoholic, tea and coffee)
	Baker's yeast
	Food additives (antioxidants, colours, flavours, stabilizers)
	Novel foods (soy sauce, tempeh, miso)
	Mushroom products
	Amino acids, vitamins
	Starch products
	Glucose and high-fructose syrups
	Functional modifications of proteins, pectins
Agriculture	Animal feedstuffs (SCP)
	Veterinary vaccines
	Ensilage and composting processes
	Microbial pesticides
	Rhizobium and other N-fixing bacterial inoculants
	Mycorrhizal inoculants
	Plant cell and tissue culture (vegetative propagation, embryo production, genetic improvement)

(Adapted from Bull *et al.*, 1982).

considerably (Fig. 3.1) medium development must be able to take advantage of choice of component ingredients.

The raw material needs of many biotechnological processes are vast and necessitate special attention to storage in order to avoid chemical changes and microbial contamination. The health of the operators must also be considered, in particular when handling powders.

Media preparation may seem to be a relatively uninteresting part of the overall process, but it is, in fact, the cornerstone of the whole operation. Poor medium design will lead to low efficiency of growth and subsequently poor product formation.

Since most fermentation processes utilize single pure cultures, all media must be presented to the fermenter vessel free of contaminating

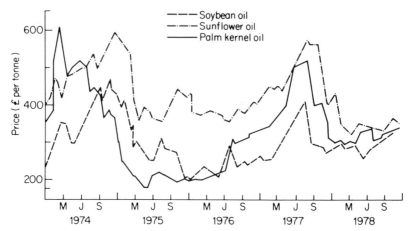

Fig. 3.1   Fluctuation in the price of some important industrial media components over the period 1974–1978.

microorganisms. The universal method for sterilizing most heterogeneous industrial media is by steam heating under pressure. Sterilization of media should be aimed at achieving the mildest treatment of the components while removing all unwanted microorganisms.

3.3   Open and closed fermenter systems

In current industrial biotechnological practice there are three main types of bioreactor and two forms of biocatalyst. Bioreactors can function on a *batch, fed-batch* (semicontinuous) or *continuous* basis. Within the bioreactor, the cultures may be static or agitated, in the presence (aerobic) or absence (anaerobic) of oxygen, and in aqueous or low moisture conditions. The biocatalysts (whole cells or enzymes) can be free or can be immobilized by attachment to surfaces. In general, the bioreactions occurring within the bioreactor will take place under moderate conditions of pH (near neutrality) and temperature (20 to 65 °C). In many bioprocesses the final products of metabolism will be present in low concentrations in an aqueous phase and will require to be separated out before commercial sale.

Bioreactor systems for organism growth may be classified as 'closed' or 'open'. A system is considered closed when an essential part of the system cannot enter or leave the system. Thus, in a traditional batch or closed fermentation system, all the nutrient components are added at the beginning of the fermentation and, as a result, the growth rate of the contained organisms will eventually decline to zero due either to diminishing nutrients or accumulation of toxic waste products. For this reason, the metabolism of organisms in closed batch processes is always in a transient or changing state. However, most current biotechnological systems function as batch processes where conditions are optimized to give the maximum formation of the required product, for example, brewing, antibiotics and enzymes.

The size of production bioreactors in industry ranges between 10 000 to 200 000 litres – although a gigantic 4 million litre bioreactor has recently been commissioned. The smaller bioreactors are principally used for high-cost, low-volume products, for example certain enzymes and chemicals, whereas the larger bioreactors are widely used to produce antibiotics, organic acids, etc.

A modification of the batch process is the fed-batch system in which volumes of nutrient may be added during the fermentation to augment depletion of nutrients or as selective activators of new compounds, for example in the production of yeast for the baking industries. However, the system remains closed since there is no continuous outflow.

In contrast to this, a fermentation system is considered open if all the components of the system (such as organisms and nutrients) can continuously enter and leave the bioreactor. Thus open or continuous flow bioreactor systems have a continuous input of fresh nutrient media and an output of biomass and other products (Fig. 3.2).

In such systems it can be expected that the rate of conversion of substrate to biomass and products will balance the output rate, and in this way a steady metabolic state will be achieved. Although continuous processes have gained wide use on a laboratory scale, few such processes have found wide industrial acceptance. However, continuous processes have been widely practised in SCP production, for example, ICI Pruteen production from methanol (Fig. 3.3) and Rank Hovis McDougall mycoprotein production.

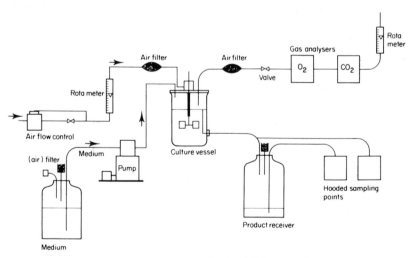

Fig. 3.2 A diagrammatic representation of a simple laboratory fermenter operating on a continuous basis.

Fig. 3.3   A giant fermenter being erected by ICI at Billingham designed to grow bacteria on methanol. The single cell product is called pruteen.

3.4   Bioreactor design

In all forms of fermentation systems the ultimate aim is to ensure that all parts of the system are subject to the same conditions. Within the bioreactor the biocatalysts are suspended in the aqueous nutrient medium containing the necessary substrates for growth of the organism and required product formation. All nutrients, including oxygen, must be provided to diffuse into each cell, and waste products such as heat, carbon dioxide and waste metabolites removed.

The concentration of the nutrients in the vicinity of the organism must be held within a definite range, since low values limit the rate of organism metabolism while excessive concentrations can be toxic. Biological reactions run most efficiently within optimum ranges of environmental parameters; in biotechnological processes these conditions must be provided on a micro scale so that each cell is equally provided for. When the large scale of many bioreactor systems is considered it will be realized how difficult it is to achieve these conditions in a whole population. It is here that the skills of the process or chemical engineer and the microbiologist must come together.

Fermentation reactions are multiphase, involving a gas phase (containing N_2, O_2 and CO_2), one or more liquid phases (aqueous medium and liquid substrate) and a solid microphase (the microorganisms and possibly solid substrates). All phases must be kept in close contact to achieve rapid mass and heat transfer. In a perfectly mixed bioreactor all reactants entering the system are immediately mixed and uniformly distributed to ensure homogeneity inside the reactor.

To design the optimum bioreactor system, the following guidelines must be closely adhered to:

(1) The bioreactor should be designed to exclude entrance of contaminating organisms as well as containing the desired organisms.
(2) The culture volume should remain constant, i.e. no leakage or evaporation.
(3) The dissolved oxygen level must be maintained above critical levels of aeration and culture agitation for aerobic organisms.
(4) Environmental parameters such as temperature, pH, etc., must be controlled.
(5) The culture volume must be well mixed.

The standards of materials used in the construction of sophisticated fermenters is also important (Table 3.2).

Fermentation technologists seek to maximize culture potential by accurate control of the bioreactor environment. There is, however, a great lack of understanding of just what environmental conditions will produce an optimal yield of organism or product.

Table 3.2 Standards of materials used in sophisticated fermenter design.

1 All materials coming into contact with the solutions entering the bioreactor or the actual organism culture must be corrosion resistant to prevent trace metal contamination of the process.

2 The materials must be nontoxic so that slight dissolution of the material or components does not inhibit culture growth.

3 The materials of the bioreactor must withstand repeated sterilization with high-pressure steam.

4 The bioreactor stirrer system, entry ports and end-plates must be easily machinable and sufficiently rigid not to be deformed or broken under mechanical stress.

5 Visual inspection of the medium and culture is advantageous; transparent materials should be used wherever possible.

To understand and control a fermentation process it is necessary to know the state of the process over a small time increment and, further, to know how the organism responds to a set of measurable environmental conditions. Process optimization therefore requires accurate and rapid feedback control. In the future, the computer will be an integral part of most fermentation processes. However, there is a lack of good senser probes that will allow online analysis to be made on the chemical components of the fermentation process.

A large worldwide market exists for the development of rapid methods of monitoring the many reactions within a bioreactor. In particular, the greatest need is for innovatory microelectronic designs.

Bioreactor configurations have changed considerably over the last few decades. The original fermentation system was a shallow tank agitated or stirred by manpower (Fig. 3.4a). From this developed the aeration tower system which now dominates industrial usage. As fermentation systems were further developed, two design solutions to the problems of aeration and agitation have been implemented. The first approach uses mechanical aeration and agitation devices, with relatively high power requirements (Fig. 3.4b); the standard example is the continuously stirred tank reactor (CSTR), widely used throughout conventional laboratory and industrial fermentations. Such bioreactors ensure good gas/liquid mass transfer, have reasonable heat transfer, and ensure good mixing of the bioreactor contents.

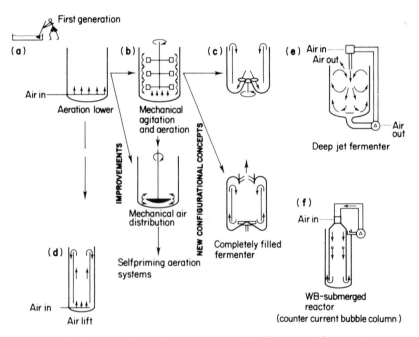

Fig. 3.4   Traditional and new fermenter designs (a). The original fermentation system (b) fermentation by mechanical aeration and agitation (c) recycle or loop fermenter (d) airlift fermenter (e) deep jet fermenter (f) WB – submerged reactor.

The second approach to bioreactor design uses air distribution (with low power consumption) to create forced and controlled liquid bulk flow in a recycle or loop bioreactor (Fig. 3.4c). In this way the contents are subjected to a controlled recycle flow, either within the bioreactor or involving an external recycle loop. Thus stirring has been replaced by pumping, which may be mechanical or pneumatic, as in the case of the airlift bioreactor (Fig. 3.4d).

The stirred tank bioreactor system is still the most widely used, but most new designs are dominated by recycle principles and several new designs have found ready acceptance because of their high efficiency in new areas of fermentation technology, for example SCP production from *n*-alkanes, methanol, and also waste-water treatment.

3.5 Scale-up

Fermentation processes are normally developed in three stages or scales. In the initial stage, basic screening procedures are carried out using relatively simple microbiological techniques, such as petri-dishes, Erhlenmeyer flasks, etc. This is followed by a pilot plant investigation to determine the optimal operating conditions in a volume capacity of 5 to 200 litres. The final stage is the transfer of the study to plant scale and final economic realization.

Throughout these stages of development the biotechnologist aims to maintain the optimal environmental conditions for the process at all levels of development. These environmental conditions involve both chemical factors (substrate concentrations, etc.) and physical factors (mass transfer ability, mixing ability, power dissipation, etc.). In particular, the physical factors create the greatest problems when moving from one scale to another. It is in this area that the chemical engineers find greatest expression of their skills.

3.6 Solid substrate fermentations

Most bioreactor studies have been concerned with liquid systems. However, mention must be made of the widely practised fermentation systems that utilize solid substrates without the presence of free water. Such solid substrate fermentations have long been practised and indeed predate liquid fermentations by several millennia. Particular examples include Oriental fermentations such as soy sauce and tempeh, mould-ripened cheeses, mushroom cultivation and composting of organic wastes (Table 3.3).

The scale of these biotechnological processes is vast but, surprisingly, tends to be overlooked by modern fermentation technologists. This may be due to the complexity of these processes and their difficulty of explanation in mathematical modelling terms. However, there has recently been a considerable increase of interest in such systems and many worthwhile studies are now in progress.

Solid substrate fermentation (SSF) is adaptable to either batch or continuous processes, and the complexity of the equipment is not very different from that

Table 3.3 Some examples of solid-state fermentation.

Example	Substrate	Microorganism(s) involved
Mushroom production (European and Oriental)	Straw, manure, wood	Agaricus bisporus Lentinus edodes Volvariella volvaceae
Soy sauce	Wheat and soybeans Soybeans	Aspergillus oryzae
Tempeh	Peanut press cake	Rhizopus spp.
Ontjom		Neurospora sitophila
Cheeses	Milk curd	Penicillium roquefortii
Leaching of metals	Low-grade ores	Thiobacillus spp.
Organic acids	Cane sugar, molasses	Aspergillus niger
Enzymes	Wheat bran, etc.	Aspergillus niger
Composting	Mixed organic material	Fungi, bacteria, actinomycetes
Sewage treatment	Components of sewage	Bacteria, fungi and protozoa

required with traditional stirred tanks. Most SSF processes involve filamentous fungi or yeast species, since such organisms can function at low moisture levels.

There is little information available in Western literature on the design of solid substrate bioreactors, although much is available in Japanese literature (in the Japanese language).

The low energy requirement of SSF processes is an obvious attraction for the development of biotechnological processes, and new studies suggest that such processes may well become increasingly exploited at an industrial level.

3.7   Technology of animal and plant cell culture

The mass cultivation of organisms for biotechnological processes has been largely developed around the bacteria, yeasts and filamentous fungi, and only more recently for plant and animal cell cultures.

The use of plant cell culture techniques for micropropagation of certain plants is discussed in Chapter 10. In these techniques plant cell cultures progress through organogenesis, plantlet amplification and eventual establishment in soil. However, large-scale production of suspension cell cultures of many plant species has now been achieved and yields of products typical of the whole plant have been impressive, e.g. nicotine, alkaloids and ginseng. It is now envisaged that large-scale fermentation programmes will be able to produce commercially acceptable levels of certain high-value products, e.g. digitalis, jasmine, spearmint, codeine, etc.

The fermentation methods used to cultivate plant cells in liquid agitated culture have been largely derived from microbial techniques. Plant cell culture is much slower than microorganism culture, though most of the other characteristics of fermentation are quite similar. The volume of an average cultured plant cell can be up to 200 000 times that of a bacteria cell. Although some plant products are now appearing on the market, the process is not expected to be commercially attractive for many years.

Many potentially important organic compounds are by necessity produced by animal or human cell cultures. Development has been hampered by the problems and cost of scale-up operations.

Animal cells can be grown either in an unattached suspension culture or attached to a solid surface. Cells such as Hela cells (cells derived from a human malignancy) can grow in either state; lymphoblastoid cells will grow in suspension culture; while primary or normal diploid cells will only grow when they are attached to a solid surface. Most future commercial development with animal cells will be dominated by the cultivation of anchorage-dependent cell types.

Monolayer cultivation of animal cells is governed by the surface area available for attachment, and design considerations have been directed to methods of increasing surface area. Early designs relied mainly on roller tubes or bottles to ensure exchange of nutrients and gases. A recent sophisticated system supports the growth of cells in coils of gas-permeable Teflon tubing, each tube having a surface area of 10 000 cm^2, and up to 20 such coils can be incorporated into an incubator chamber. A wide range of cells has been successfully cultured under these conditions.

Suspension cultures have been successfully developed to quite large bioreactor volumes, thus utilizing all the engineering advantages of the stirred tank bioreactor that have accrued from microbial studies. Such studies have only been carried out on a batch culture basis.

A combination of attachment culture and suspension culture by the use of *microcarrier beads* has been a major innovation in this area. In principle, the anchorage-dependent cells attach to special DEAE-Sephadex beads (having a surface area of 7 $cm^2 mg^{-1}$) which are able to float in suspension. In this way the engineering advantages of the stirred bioreactor may be used with anchored cells. Many cell types have been grown in this manner, with successful production of viruses and human interferon. The undoubted success of the microcarrier beads may eventually lead to the demise of conventional monolayer systems. New bioreactor designs involving the microcarrier bead concept will surely create a wider commercial development of animal and human cell types.

3.8 Fermentation technology in developing countries

In industrialized countries bioreactor design has largely tended towards extremely expensive stainless steel or copper reactor vessels. Cheaper alternatives such as wood, plastic or concrete can be used without too much impairment of many fermentation processes. Hitherto, the fermentation processes in developed countries have been largely aerobic processes necessitating highly expensive design considerations (high technology) to ensure adequate oxygen transfer. Greater attempts should be made to develop anaerobic fermentations (low technology) for the production of essential products.

Industrial microbiology has been further dominated by pure culture requirements. The advantages of such systems cannot be denied, but they involve costly sterilization methods and skilled operator control. Many fermentation systems could work quite successfully with low levels of contaminating microbes, thereby reducing running costs and requiring less skilled operators.

Finally, modern bioreactor technology relies on sophisticated instrumentation for monitoring purposes. In part, this has come about because of the ever-increasing cost of skilled labour. Developing countries, however, have an abundance of manpower, and this could be profitably used to watch and adjust ongoing systems.

3.9 Downstream processing

It is not just enough to grow the required cells in a bioreactor; extraction and purification of the desired end product ('downstream processing') from the bioprocess is equally – if not more – important, calling on the skills of chemists and chemical engineers as well as bioscientists and process engineers.

The design and efficient operation of downstream processing operations are vital elements in getting the required products into commercial use and should reflect the need not to lose more of the desired product than is absolutely necessary. An example of the effort expended in downstream processing is provided by the new plant Eli Lilly built to produce human insulin (Humulin). Over 90% of the 200 staff are involved in recovery processes. Thus, downstream processing represents a major part of the overall costs of most biotechnological processes, although it is the least glamorous aspect of biotechnology. Improvements in downstream processing will improve the overall efficiency and costs of processes.

Downstream processing is primarily concerned with initial separation of the bioreactor broth into a liquid phase and a solids phase, and subsequent concentration and purification of the product. Processing normally involves more than one stage (Table 3.4). Methods in use (or proposed) range from the conventional to the arcane, including distillation, centrifuging, filtration, ultrafiltration, solvent extraction, adsorption, selective membrane technology, reverse osmosis, molecular sieves, electrophoresis and affinity chromatography (Fig. 3.5). It is in this area that several potential industrial applications of modern biotechnology have come to grief, either because the extraction has defeated the ingenuity of the designers, or – more probably – because the extraction process required so much energy input that it was uneconomic.

Final products of the downstream purification stages should have some degree of stability for commercial distribution. Stability is best achieved for most products by using some form of drying. In practice this is achieved by spray drying, fluidized bed drying or by freeze drying. The method of choice is product and cost dependent. Products sold in the dry form include organic acids, amino acids, antibiotics, polysaccharides, enzymes, SCP

Table 3.4 Downstream processing operations.

Separation	Filtration
	Centrifugation
	Flotation
	Disruption
Concentration	Solubilization
	Extraction
	Thermal processing
	Membrane filtration
	Precipitation
Purification	Crystallization
	Chromatography
Modification	
Drying	

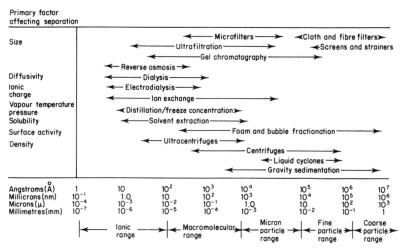

Fig. 3.5 Ranges of applications of various unit operations. Reproduced by permission from Atkinson and Marituna (1983).

and many others. Many products cannot be supplied easily in a dried form and must be sold in liquid preparations. Care must be taken to avoid microbial contamination and deterioration, and (when the product is proteinaceous) denaturation.

Downstream processing will continue to be the most challenging and demanding part of many biotechnological processes. Purity and stability are the hallmarks of most high-value biotechnological products.

In conclusion, it can be said that biotechnological processes usually need to be contained within a defined area or system, and to a large extent the ultimate success of most processes depends on the correct choice and operation of these systems. On the industrial side, the scale of operation will, for economic reasons, mainly be very large, and in almost all cases the final success will require the closest cooperation between the bioscientist and the process engineer – demonstrating the truly interdisciplinary nature of biotechnological processes.

4

Genetics and Biotechnology

4.1 Introduction

In essence, all properties of organisms depend on the sum of their genes. There are two broad categories of genes – structural and regulatory. Structural genes encode for amino acid sequences of proteins which, as enzymes, determine the biochemical capabilities of the organism by catalysing particular synthetic or catabolic reactions, or, alternatively, play more static roles as components of cellular structures. In contrast, the regulatory genes control the expression of the structural genes by determining the rate of production of their protein products in response to intracellular or extracellular signals. The derivation of these principles has been achieved using well-known genetic techniques which are not considered further here.

The linking of DNA to genes and heredity was recognized in the 1940s and since then there has been a spectacular unravelling of the complex interactions required to translate the coded chemical information of the DNA molecule into cellular and organismal expression. Change in the DNA molecule making up the genetic complement of an organism is the means by which organisms evolve and adapt to new environments. In nature, changes in the DNA of an organism can occur in two ways:

(1) By *mutation*, which is a chemical deletion or addition of one or more of the chemical parts of the DNA molecule.
(2) By the interchange of genetic information or DNA between like organisms, normally by *sexual reproduction*. This is achieved by a process of *conjugation* in which there is a donor, called male, and a recipient, called female.

Classical genetics was until recently the only way in which heredity could be studied and manipulated. However, in recent years, new techniques have permitted unprecedented alterations in the genetic make-up of organisms, even allowing exchange of DNA between unlike organisms.

The manipulation of the genetic material in organisms can now be achieved in three clearly definable ways – organismal, cellular and molecular.

4.1.1   Organismal

Genetic manipulation of whole organisms has been happening naturally by sexual reproduction since the beginning of time. The evolutionary progress of almost all living creatures has involved active interaction between the genomes and the environment. Active control of sexual reproduction in plants and animals has been practised in agriculture for decades – even centuries. In more recent times it has been used with several industrial microorganisms, e.g. yeasts. It involves selection, mutation, sexual crosses, hybridization, etc. However, it is a very random process and can take a long time to achieve desired results – if at all in some cases. In agriculture, the benefits have been immense with much improved plants and animals, while in the biotechnological industries there have been greatly improved productivies, e.g. antibiotics, enzymes, etc.

4.1.2   Cellular

Cellular manipulations of DNA have been used for nearly two decades, and involve either cell fusion or the culture of cells and the regeneration of whole plants from these cells (see Chapter 10). This is a semi-random or directed process, in contrast to organismal manipulations, and the changes can be more readily identified. Successful biotechnological examples of these methods include monoclonal antibodies (see Section 4.3) and the cloning of many important plant species such as oil palms, tomatoes, etc.

4.1.3   Molecular

Molecular manipulations of DNA and RNA are little over a decade old. They heralded a new era of genetic manipulations, enabling – for the first time in biological history – directed control of changes in organisms. This is the much publicized area of *genetic engineering* or *recombinant DNA technology* which is confidently believed will bring dramatic changes to biotechnology. In these techniques the experimenter is able to know much more about the genetic changes being made. It is now possible to add or delete parts of the DNA molecule with a high degree of precision and the product can be easily identified. Current industrial ventures are concerned with the production of numerous compounds ranging from pharmaceuticals to commodity chemicals, and are discussed in later chapters.

4.2   Industrial genetics

Biotechnology has so far been considered as an interplay between two components, one of which is the selection of the best biocatalyst for a particular process, while the other is the construction and operation of the best environment for the catalyst to achieve optimum operation.

The most effective, stable and convenient form for the biocatalyst is a whole organism; in most cases it is some type of microbe, for example bacterium, yeast or mould, although animal cell cultures and (to a lesser extent) plant cell cultures are finding ever increasing uses in biotechnology.

Microorganisms used in biotechnological processes were originally isolated from the natural environment, and have subsequently been modified by the industrial geneticist into superior organisms for specific productivity. The success of strain selection and improvement programmes practised by all biologically based industries (e.g. brewing, antibiotics, etc.) is a direct result of the close cooperation between the technologist and the geneticist. In the future, this relationship will be even more necessary in formulating the specific physiological and biochemical characteristics that are sought in new organisms in order to give the fullest range of biological activities to biotechnology.

In biotechnological processes, the aim is to optimize the particular characteristics sought in an organism, for example specific enzyme production, by-product formation, etc. Genetic modification to improve productivity has been widely practised. The task of improving yields of some primary metabolites and macromolecules (e.g. enzymes) is simpler than trying to improve the yields of complex products such as antibiotics.

Advances have been achieved in this area by using *screening* and *selection* techniques to obtain better organisms. In a selection system all rare strains grow while the rest do not; in a screening system all strains grow, but certain strains or cultures are chosen because they show the desired qualities required by the industry in question.

In most industrial genetics the basis for changing the organism's genome has been by mutation using X-rays and mutagenic chemicals. However, such methods normally lead only to the loss of undesired characters or increased production due to loss of control functions. It has rarely led to the appearance of a new function or property. Thus, an organism with a desired feature will be selected from the natural environment, propagated and subjected to a mutational programme; then screened to select the best progeny.

Unfortunately, many of the microorganisms that have gained industrial importance do not have a clearly defined sexual cycle. In particular, this has been the case in antibiotic-producing microorganisms; this has meant that the only way to change the genome with a view to enhancing productivity has been to indulge in massive mutational programmes followed by screening and selection to detect the new variants that might arise.

Once a high-producing strain has been found, great care is required in maintaining the strain. Undesired spontaneous mutations can sometimes occur at a high rate, giving rise to degeneration of the strain's industrial potential. Strain instability is a constant problem in industrial utilization of microorganisms. Industry has always placed great emphasis on strain viability and productivity potential of the preserved biological material. Most industrially important microorganisms can be stored for long periods, for example in liquid nitrogen, by lyophilization (freeze drying) or under oil, and still retain their desired biological properties.

However, despite elaborate preservation and propagation methods, a strain has generally to be grown in a large production bioreactor in which the chances of genetic changes through spontaneous mutation and selection are very high. The chance of a high rate of spontaneous mutation is probably greater when the industrial strains in use have resulted from many years of mutagen treatment.

Great secrecy surrounds the use of industrial microorganisms and immense care is taken to ensure that they do not unwittingly pass to outside agencies.

There is now a growing movement away from the extreme empiricism that characterized the early days of the fermentation industries. Fundamental studies of the genetics of microorganisms now provide a background of knowledge for the experimental solution of industrial problems and have contributed directly to progress in industrial strain selection.

In recent years, industrial genetics has come to depend increasingly on the two new ways of manipulating DNA – protoplast and cell fusion, and recombinant DNA technology. These are now important additions to the technical repertoire of the geneticists involved with biotechnological industries. A brief examination of these techniques attempts to show their modern relevance to biotechnology.

4.3   Protoplast and cell fusion technologies

Plant and most microbial cells are characterized by a distinct outer wall or exoskeleton which gives the shape characteristic to the cell or organism. Immediately within the cell wall is the living membrane or plasma membrane, retaining all the cellular components such as nuclei, mitochondria, vesicles, etc. For some years now it has been possible, using special techniques (in particular hydrolytic enzymes), to remove the cell wall, releasing spherical membrane-bound structures known as protoplasts. These protoplasts are extremely fragile but can be maintained in isolation for variable periods of time. Isolated protoplasts cannot propagate themselves as such, requiring first to regenerate a cell wall before regaining reproductive capacity.

In practice, it is the cell wall that largely hinders the sexual conjugation of unlike organisms. Only with completely sexually compatible strains does the wall degenerate allowing protoplasmic interchange. Thus natural sexual mating barriers in microorganisms may, in part, be due to cell wall limitations, and by removing this cell wall, the likelihood of many cellular fusions will increase.

Protoplasts can be routinely obtained from many plant species, bacteria, yeasts and filamentous fungi. Protoplasts from different strains can sometimes be persuaded to fuse and so overcome the natural sexual mating barriers. However, the range of protoplast fusions is severely limited by the need for DNA compatibility between the strains concerned. Fusion of protoplasts can be enhanced by treatment with the chemical polyethylene glycol which, under optimum conditions, can lead to extremely high frequencies of recombinant formation, which can be increased still further by ultraviolet irradiation of the parental protoplast preparations. Protoplast fusion can also occur with human or animal cell types.

Protoplast fusion has obvious empirical applications in yield improvement of antibiotics by combining yield-enhancing mutations from different strains or even species. Protoplasts are also an important part of genetic engineering, in facilitating recombinant DNA transfer. In the long term, fusion may provide a method of reassorting whole groups of genes between different strains of macro- and micro-organisms. Many exciting new areas of research are present in this field, including a claim that a line of human cells capable of secreting insulin and growing in suspension cultures has been generated by cell fusion.

One of the most exciting and commercially rewarding areas of biotechnology involves a form of animal cell fusion leading to the formation of monoclonal antibodies. It has long been recognized that certain cells (β-lymphocytes) within the body of vertebrates have the ability to secrete antibodies (see Chapter 9) which can inactivate contaminating or foreign molecules (the antigen) within the animal system. Furthermore, it is now known that individual β-lymphocyte cells produce single antibody types. Attempts to cultivate the antibody-producing cells in artificial media generally proved unsuccessful, with the cells either dying or ceasing to produce the antibodies.

However, in 1975 George Kohler and Cesar Milstein successfully demonstrated the production of pure or monoclonal antibodies from the fusion product (*hybridoma*) of β-lymphocytes (antibody-producing cells) and myeloma tumour cells. In 1984 they were awarded a Nobel prize for this outstanding scientific achievement.

The monoclonal antibody technique changes antibody-secreting cells (with limited life span) into cells capable of continuous growth while maintaining their specific antibody-secreting potential. This 'immortalization' is achieved by a fusion technique, whereby β-lymphocyte cells are fused to 'immortal' cancer or myeloma cells in a one-to-one ratio, forming hybrids or hybridomas capable of continuous growth and antibody secretion in culture. Single hybrid cells can then be selected and grown as clones or pure cultures of the hybridomas. Such cells continue to secrete antibody; the antibody is of one particular specificity, as opposed to the mixture of antibodies that occurs in an animal bloodstream after conventional methods of immunization.

Monoclonal antibody formation is performed by injecting a mouse or rabbit with the antigen, later removing the spleen and then allowing fusion of individual spleen cells with individual myeloma cells. Approximately 1% of the spleen cells and 10% of the final hybridomas are antibody-secreting cells (Fig. 4.1). Techniques are available to identify the right antibody-secreting hybridoma cell, cloning or propagating that cell into large populations with subsequent formation of quantities of the desired antibody. These cells may be frozen and later reused.

Monoclonal antibodies have gained wide application in many diagnostic techniques requiring a high degree of specificity. Specific monoclonal antibodies have been combined into test kits for diagnostic purposes, e.g. for breast and prostate cancer, hepatitis B virus, microbial toxins, etc. Monoclonal antibodies may also be used in the future to carry cytotoxic drugs to the site of cancer cells, while in the fermentation industry they are already widely used as affinity ligands to bind and purify expensive products such as the interferons.

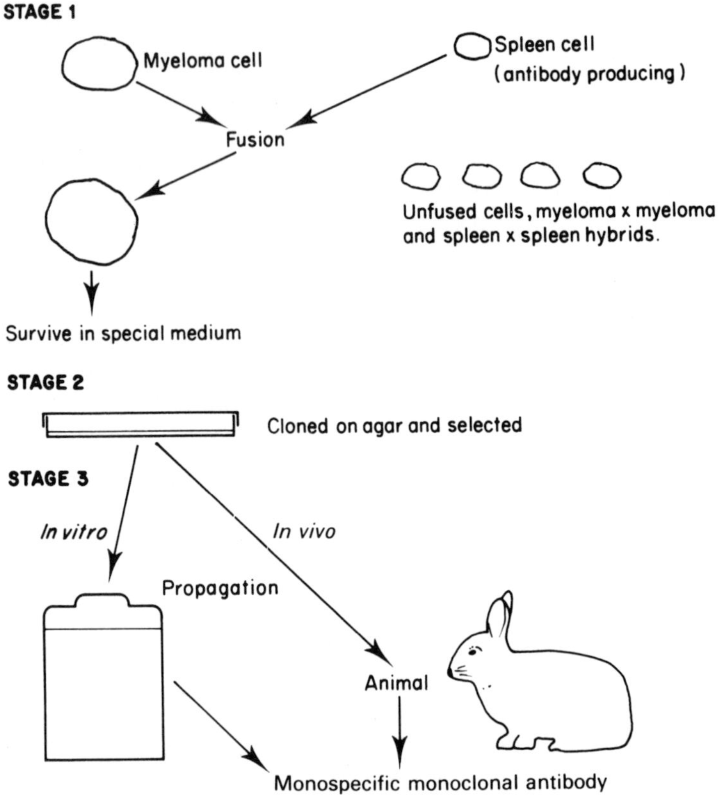

STAGE 1

Myeloma cell

Spleen cell
(antibody producing)

Fusion

Unfused cells, myeloma x myeloma
and spleen x spleen hybrids.

Survive in special medium

STAGE 2

Cloned on agar and selected

STAGE 3

In vitro   *In vivo*

Propagation

Animal

Monospecific monoclonal antibody

Fig. 4.1   Diagram illustrating the formation of antibody-producing hybridomas by the fusion technique. Stage 1, myeloma cells and antibody producing cells (derived from an immunized animal or man) are incubated in a special medium containing polyethylene-glycol which enhances fusion. Stage 2, the myeloma spleen hybridoma cells are selected out and cultured in closed agar dishes. Stage 3, the specific antibody producing hybridoma is selected and propogated in culture vessels (*in vitro*) or in animal (*in viro*) and mono-clonal antibodies harvested.

In the ten or so years since the development of the first monoclonal antibody the methodology has developed from a purely scientific tool into one of the fastest expanding fields of biotechnology – in particular supplying a whole new range of improved diagnostic test kits. The monoclonal antibody market is expected to continue to grow at a very high rate, and in health care alone the anticipated world market is between $800 million and $2700 million by the year 2000.

4.4 Genetic engineering

Genetic recombination, as occurs during normal sexual reproduction, consists of the breakage and rejoining of DNA molecules of the chromosomes, and is of fundamental importance to living organisms. The naturally occurring mechanisms that permit genetic recombination to take place are generally considered to be confined by strong taxonomic constraints.

In contrast, recombinant DNA techniques, popularly termed *gene cloning* or *genetic engineering*, offer potentially unlimited opportunities for creating new combinations of genes which at the moment do not exist under natural conditions.

Genetic engineering has been defined as 'the formation of new combinations of heritable material by the insertion of nucleic acid molecules, produced by whatever means outside the cell, into any virus, bacterial plasmid or other vector system so as to allow their incorporation into a host organism in which they do not naturally occur but in which they are capable of continued propagation'.

These techniques allow the splicing of DNA molecules of quite diverse origin and, when combined with techniques of genetic transformation, etc., facilitate the introduction of foreign DNA into other organisms, particularly bacteria.

Thus DNA can be isolated from cells of plants, animals or microorganisms (the donors) and can be fragmented into groups of one or more genes. Such fragments can then be coupled to another piece of DNA (the vector) and then passed into the host or recipient cell, becoming part of the genetic complement of the new host. The host cell can then be propagated in mass to provide novel genetic properties and chemical abilities that were unattainable by conventional ways of selective breeding or mutation. Although much work to date has involved bacteria, the techniques are evolving at an astonishing rate and ways have been developed for introducing DNA into other organisms such as yeasts and plant and animal cell cultures. Provided that the genetic material transferred in this manner can replicate and be expressed in the new cell type, there are virtually no limits to the range of organisms with new properties that could be produced by genetic engineering.

These methods potentially allow totally new functions to be added to the capabilities of industrial microorganisms, and open up vistas for the genetic engineering of industrial microorganisms, together with animal and plant cell cultures, that are quite breathtaking in their scope. This is undoubtedly one of the most important areas of development in biotechnology; it places within reach processes for the production of simple and complex chemical substances hitherto deemed impracticable by microbial manipulation. Examples include the synthesis in microorganisms of specific animal proteins such as insulin, enhanced ranges of enzymes, hormones, antitumour and antiviral compounds (interferon), fine chemicals or bulk chemicals such as ethanol, or the ability to utilize complex substrates such as cellulose and lignin and to derive worthwhile products from them.

Genetic engineering holds the potential to extend the range and power of every aspect of biotechnology. In the first instance these techniques will be widely used to improve existing microbial processes, by improving stability of existing cultures and eliminating unwanted side-products. It is confidently anticipated that within this decade recombinant DNA techniques will have established new microorganisms with new and unusual metabolic properties. Fermentations based on these technical advances could become competitive with petrochemicals for producing a whole range of chemical compounds, for example ethylene glycol (used in the plastics industry). In the food industry, improved strains of bacteria and fungi will soon be influencing such traditional processes as baking and cheesemaking, bringing greater control and reproducibility of flavour and texture.

To understand the working concepts of recombinant DNA technology requires a good knowledge of molecular biology. A brief explanation is given here, but readers are advised to consult some of the many excellent texts that are available in this field (e.g. Old and Primrose, 1980).

The basic molecular requirements for the *in vitro* transfer and expression of foreign DNA in a host cell (gene transfer technology) are as follows:

(1)　*The vector or carrier system.* Two broad categories of vector molecules have been developed as vehicles for gene transfer – plasmids (small units of DNA distinct from chromosomes) and bacteriophages (or bacterial viruses). Vector molecules will normally exist within a cell in an independent or extrachromosomal form, not becoming part of the chromosomal system of the organism. Vector molecules should be capable of entering the host cell and replicating within it. Ideally, the vector should be small, easily prepared and must contain at least one site where integration of foreign DNA will not destroy an essential function. Plasmids undoubtedly offer the greatest potential in biotechnology and are found in an increasingly wide range of organisms, for example, bacteria, yeasts and mould fungi; they have been mostly studied in gram-negative bacteria.

(2)　*Splicing genes.* The most significant advances towards the construction of hybrid DNA molecules *in vitro* have come from the discovery that site-specific *restriction endonuclease enzymes* produce specific DNA fragments that can be joined to any similarly treated DNA molecule using another enzyme, *DNA ligase.* Restriction enzymes are present in a wide range of bacteria and can distinguish between DNA from their own cells and foreign DNA by recognizing certain sequence of nucleotides. There are techniques available for breaking open a length of DNA into shorter fragments, which contain a number of genes determined by the enzyme used. Such DNA fragments can then be separated from each other on the basis of differing molecular weights, and can subsequently be joined together in a number of ways, provided that the ends are complementary. The sources of DNA can be quite different, giving an opportunity to replicate the DNA biologically by inserting it into other cells.

The composite molecules in which DNA has been inserted have been termed 'DNA chimaeras' because of the analogy with the chimaera of mythology, a creature with the head of a lion, the body of a goat and the tail of a serpent.

(3) *Introduction of vector DNA recombinants.* The new recombinant DNA can now be introduced into the host cell by *transformations* (the direct uptake of DNA by a cell from its environment) or *transductions* (DNA transferred from one organism to another by way of a carrier or vector system) and if acceptable the new DNA will be cloned with the propagation of the host cell.

(4) *Assaying foreign gene products.* Many methods are now available for the detection of hybrid DNA.

The strategies involved in genetic engineering are outlined in Table 4.1 and Fig. 4.2.

Although the theory underlying the exchange of genetic information between unrelated organisms and their propagation in bioreactors is becoming better

Table 4.1 Strategies involved in genetic engineering.

Formation of DNA fragments	Extracted DNA can be cut into small sequences by specific enzymes, restriction endonucleases found in many species of bacteria.
Splicing of DNA into vectors	The small sequences of DNA can be joined or spliced into the vector DNA molecules by an enzyme DNA ligase creating an artificial DNA molecule.
Introduction of vectors into host cells	The vectors are either viruses or plasmids, and are replicons and can exist in an extrachromosomal state; transfer normally by transduction or transformation.
Selection of newly acquired DNA	Selection and ultimate characterization of the recombinant clone.

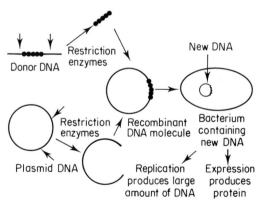

Fig. 4.2 Recombinant DNA: the technique of recombining genes from one species with those of another.

understood, difficulties still persist at the level of application and realization. Further research is required before such exchanges become commonplace and the host organisms propagated in large quantities.

To date, most genetic engineering has been accomplished with the ubiquitous gram-negative bacterium *Escherichia coli*. The wide range of available plasmid and bacteriophage vectors for *E. coli* is in marked contrast to the availability of vectors for other organisms. With mammalian cells there is only one potential vector, the simian virus (SV40), and some evidence of foreign gene expression has been achieved with monkey cells. The *Agrobacterium* system is being extensively and successfully used as a vector in plant systems.

The ability to clone DNA and to study its expression in controlled environments is a development of quite unparalleled importance in modern bioscience. The further expression of genetic engineering in the context of biotechnology, where newly created gene pools in organisms can be expressed in large quantities, will offer unbounded opportunities for the well-being of mankind.

4.5   Potential risks of genetic engineering

The early studies on gene manipulation provoked wide discussion and considerable concern at the possible risks that could arise with certain types of experiment. It was believed by some that the construction of recombinant DNA molecules and their insertion into microorganisms could create novel organisms which might inadvertently be released from the laboratory and become a biohazard to humans or the environment. In contrast, others considered that newly synthesized organisms with their additional genetic material would not be able to compete with the normal strains present in nature.

The present views of gene manipulation studies are becoming more moderate as experiments have shown that this work can proceed within a strict safety code involving physical and biological containment of the organism.

The standards of containment enforced in the early years of recombinant DNA studies were unnecessarily restrictive and there has been a steady relaxation of the regulations governing much of the routine genetic engineering activities. However, for many types of study, particularly with already pathogenic microorganisms, the standards will remain stringent.

Thus, for strict physical containment, laboratories involved in this type of study must have highly skilled personnel and correct physical containment equipment, for example negative pressure laboratories, autocloves, safety cabinets, etc.

Biological containment can be achieved or enhanced by selecting nonpathogenic organisms as the cloning agents of foreign DNA, or by the deliberate genetic manipulation of a microorganism to reduce the probability of survival and propagation in the environment. *E. coli*, a bacterium that is extremely prevalent in the intestinal tracts of warm-blooded and cold-blooded animals as well as in humans is the most widely used cloning agent. To offset the risk of this cloning agent becoming a danger in the environment, a special strain of

E. coli has been constructed by genetic manipulation which incorporates many fail-safe features. This strain can only grow under special laboratory conditions and there is no possibility that it can constitute a biohazard if it escapes from the laboratory.

The government-controlled Health and Safety Executive controls and monitors recombinant DNA work within the UK. This committee seeks advice from the Genetic Manipulation Advisory Group (GMAG) who formulate realistic procedural guidelines which, in general, have proved acceptable to the experimenting scientific community. Most other advanced scientific nations involved in recombinant DNA studies have set up similar advisory committees.

5

Enzyme Technology

5.1 The nature of enzymes

Enzymes are complex organic molecules present in living cells, where they act as catalysts bringing about chemical changes in substances. With the development of the science of biochemistry has come a fuller understanding of the wide range of enzymes present in living cells and of their mode of action. Without enzymes there can be no life. Although enzymes are only formed in living cells many can be separated from the cells and can continue to function *in vitro*. This unique ability of enzymes to perform their specific chemical transformations in isolation has led to an ever-increasing use of enzymes in industrial processes, collectively termed enzyme technology (see Table 5.1).

Enzyme technology embraces production, isolation, purification, use in soluble form and finally the immobilization and use of enzymes in a wide range of reactor systems. Enzyme technology will undoubtedly contribute to the solution of some of the most pressing problems modern-day society is confronted with – for example, food production, energy shortage and preservation, and improvement of the environment. This new technology has its origins in biochemistry but has drawn heavily on microbiology, chemistry and process engineering to achieve the present status of the science. For the future, enzyme technology and genetic engineering will be very closely related areas of study, dealing with the application of genes and of their products. Together

Table 5.1 Approximate annual world production of industrial enzymes.

Enzyme	Tonnes pure enzyme
Bacillus protease	500
Amyloglucosidase	300
Bacillus amylase	300
Glucose isomerase	50
Microbial rennet	20
Fungal amylase	20
Pectinase	20
Fungal protease	10

these sciences will attempt to exploit the continuous flow of discoveries being made by molecular geneticists and enzymologists.

5.2 The application of enzymes

For thousands of years processes such as brewing, breadmaking and production of cheese have involved the unrecognized use of enzymes (see Table 5.2). In this way, traditional practices and technologies that relied on enzymic conversions were well established before any coherent body of knowledge on their rational application had been developed.

In the West the industrial understanding of enzymes revolved around yeast and malt at a time when the traditional brewing and baking industries were rapidly expanding. Much of the early development of biochemistry was centred around yeast fermentations and processes for conversion of starch to sugar. In the Orient the comparable industries were saki production and various food fermentations, all of which made use of filamentous fungi as the source of enzyme activity. The year 1896 saw the true beginnings of modern microbial enzyme technology with the first marketing in the west of *takadiastase*, a rather crude mixture of hydrolytic enzymes prepared by growing the fungus *Aspergillus oryzae* on wheat bran. The method of takadiastase production varied little from that practised for thousands of years in Asia, but it did represent an important transfer of technology from East to West.

It was not until the mid-1950s that rapid development in enzyme technology occurred, using, in particular, microbial enzyme sources. The reasons for this were chiefly as follows:

(1) There was a major development in submerged cultivation practices with microorganisms primarily associated with antibiotic production, and this newly acquired knowledge was readily applied to the production of microbial enzymes.
(2) Basic knowledge of enzyme properties was rapidly expanding, and this led to the realization of the potential for using enzymes as industrial catalysts.
(3) Most enzymes of potential industrial importance could be produced from microorganisms.

The further development of enzyme additives was largely to provide enhancement of traditional processes rather than to open up new possibilities. Even at the present time, most bulk production of crude enzymes is concerned largely with enzymes that hydrolyse the glucosidic links of carbohydrates such as starch and pectins, and with the proteases which hydrolyse the peptide links of proteins.

Cell-free enzymes have many advantages over chemical processes where a number of sequential reactions are involved. In fermentation processes the use of microbial cells as catalysts can have a number of limitations:

Table 5.2 Industrial applications of enzymes.

Application	Enzymes used	Uses	Problems
Biological detergents	Primarily proteinases, produced in an extracellular form from bacteria	Used for pre-soak conditions and direct liquid applications	Allergic response of process workers; now overcome by encapsulation techniques
	Amylase enzymes	Detergents for machine dishwashing to remove resistant starch residues	
Baking industry	Fungal alpha-amylase enzymes; normally inactivated about 50 °C, destroyed during baking process	Catalyse breakdown of starch in the flour to sugar which can be used by the yeast. Used in production of white bread, buns, rolls	
	Proteinase enzymes	Biscuit manufacture to lower the protein level of the flour	
Brewing industry	Enzymes produced from barley during mashing stage of beer production	Degrade starch and proteins to produce simple sugars, amino acids and peptides used by the yeasts to enhance alcohol production	
	Industrially produced enzymes:	Now widely used in the brewing process:	
	amylases, glucanases, proteinases	split polysaccharides and proteins in the malt	
	betaglucanase	improve filtration characteristics	
	amyloglucosidase	low-calorie beer	
	proteinases	remove cloudiness during storage of beers	

Industry	Enzyme	Application	Notes
Dairy industry	Rennin, derived from the stomachs of young ruminant animals (calves, lambs, kids)	Manufacture of cheese, used to split protein	Older animals cannot be used as with increasing age rennin production decreases and is replaced by another proteinase, pepsin, which is not suitable for cheese production. In recent years the great increase in cheese consumption together with increased beef production has resulted in increasing shortage of rennin and escalating prices
	Microbially produced enzyme	Now finding increasing use in the dairy industry	
	Lipases	Enhance ripening of blue-mould cheeses (Danish Blue, Roquefort)	
	Lactases	Break down lactose to glucose and galactose	
Starch industry	Amylases, amyloglucosidases and glucoamylases	Convert starch into glucose and various syrups	
	Glucose isomerase	Converts glucose into fructose (high-fructose syrups derived from starchy materials have enhanced sweetening properties and lower calorific values)	
	Immobilized enzymes	Production of high fructose syrups	Widely used in USA and Japan but EEC restrictive practices to protect sugar beet farmers prohibits use

Table 5.2 *Contd*

Application	Enzymes used	Uses	Problems
Textile industry	Amylase enzymes	Now widely used to remove starch which is used as an adhesive or size on threads of certain fabrics to prevent damage during weaving. (Traditionally, desizing using strong chemicals has prevailed)	
	Bacterial enzymes	Generally preferred for desizing since they are able to withstand working temperatures up to 105–110 °C	
Leather industry	Enzymes found in dog and pigeon dung	Traditionally used to treat leather to make it pliable by removing certain protein components. (The process is called bating; strong bating required to achieve a soft, pliable leather, slight bating for the soles of shoes)	Offensive preparation
	Trypsin enzymes from slaughterhouses and from microorganisms	Now largely replacing the enzymes mentioned above for bating. Also used for removing the hair from hides and skins	
Medical and pharmaceutical uses	Trypsin	Debridement of wounds, dissolving blood clots	
	Pancreatic trypsin	Digestive aid formulations, treatment of inflammations etc.	
		Many enzymes used in clinical chemistry as diagnostic tools	

(1) A high proportion of the substrate will normally be converted to biomass.
(2) Wasteful side reactions may occur.
(3) The conditions for growth of the organisms may not be the same for product formation.
(4) The isolation and purification of the desired product from the fermentation liquor may be difficult.

Many, if not all, of these limitations may be alleviated by the use of purified enzymes and possibly by the further use of enzymes in an immobilized form. In the future many traditional fermentations may be replaced by multienzyme reactors that would create highly efficient rates of substrate utilization, higher yields and higher product uniformity.

There is now a rapid proliferation of uses and potential uses for more highly purified enzyme preparations in industrial processing, clinical medicine and laboratory practice. The range of pure enzymes now available commercially is rapidly increasing. The use of such highly refined preparations is largely related to cost, which can be as much as hundreds of pounds per gram. In contrast, the crude enzymes, which may have less than 1% active material, may be costed in pounds per kilogram. In many operations, such as clarifying wines and juices, chill proofing of beer and improving bread doughs, the use of crude enzymes is likely to add very little to the cost of the product.

Most of the enzymes used on an industrial scale are extracellular enzymes, i.e. enzymes that are normally excreted by the microorganisms to act upon their substrate in an external environment, and are analogous to the digestive enzymes of man and animals. Thus, when microorganisms produce enzymes to split large external molecules into an assimilable form, the enzymes are usually excreted into the fermentation medium. In this way the fermentation broth from the cultivation of certain microorganisms, for example, bacteria, yeasts or filamentous fungi, becomes a major source of proteases, amylases and (to a lesser extent) cellulases, lipases, etc. Most industrial enzymes are hydrolases and are capable of acting without complex cofactors; they are readily separated from microorganisms without rupturing the cell walls, and are water-soluble.

Some intracellular enzymes are now being produced industrially and include glucose oxidase for food preservation, asparaginase in cancer therapy and penicillin acylase for antibiotic conversion. Since most cellular enzymes are by nature intracellular, more advances can be expected in this area.

The sales of industrial enzymes were relatively small until about 1965 (Fig. 5.1) when enzymes in detergents came into general use. There was a massive increase in production over the next few years, but this was to collapse when allergic symptoms were discovered in workers handling enzymes at the factory level. There was much public concern and enzymes were mostly taken out of detergents. However, with proper precautions in the factories and by encapsulating the enzymes before they reached the customer, most risks were eliminated. Once again the application of enzymes in detergents has achieved good levels, and there is a steady growth in the use of enzymes in that part of the detergent industry where enzymes can improve washing results. In the starch industry, amylase and amyloglucosidase have substituted for acid completely in the manufacture of dextrose.

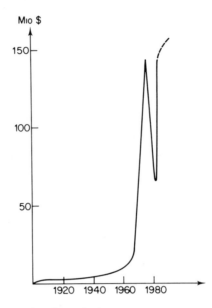

Fig. 5.1 ,World-wide sales of microbial enzymes.

Enzyme prices have fallen in real terms over the past decade. For example, the bulk quantities of enzymes for most food applications are now at least in relative terms 20 to 35% cheaper than in the mid-1970s. More specialized enzymes, used in smaller concentrations and in higher purities, have increased in use because of improved production methods. Further large-scale uses of enzymes as catalysts will be achieved only if their costs continue to fall. Current sales of industrial enzymes worldwide are between $580 million and $650 million, according to the US Department of Commerce. In financial terms, 80% of the industrial enzyme sales goes to three principal markets – starch conversion (40%), detergents (30%), and dairy applications, particularly rennets (10%). Animal rennet sales for cheese manufacturing are approaching $100 million, and will soon be affected by microbial rennets. However, the growth of enzyme sales has been and continues to be heavily influenced by the starch and detergent industries.

New innovations such as recombinant DNA technologies, and improved fermentation methods and downstream processing, will increasingly reduce production costs, particularly of high-cost enzymes, making them more competitive with other chemical methods.

Although many specific enzymes are increasingly used in clinical or diagnostic applications, the amount of enzyme actually needed is quite small. This arises from the development of automated procedures which use immobilized enzymes and seek to miniaturize the system, with the enzyme becoming analogous to the microchip in a computer. Thus, although the enzyme is essential the market need is quite small.

Table 5.3 Production of industrial enzymes by tonnage in Western world.

Nation	Tonnage (tonnes)	%
USA	6 360	12
Japan	4 240	8
Denmark	24 910	47
France	1 590	3
Germany (West)	3 180	6
Netherlands	10 070	19
UK	1 060	2
EEC	40 810	77
Switzerland	1 060	2
Others	530	1
Total	53 000	100

(From Towalski and Rothman, in *The Biotechnology Challenge*, 1986).

When enzymes are used as bulk additives, only 1 or 2 kilograms will normally be required to react with 1000 kilograms of substrate. In this way the cost of the enzyme will be between $3 to $25 per kilogram, or 10 to 14% of the value of the end product. Such enzymes are usually sold in liquid formulations and are rarely purified. In contrast, diagnostic enzymes will generally be used in milligram or microgram quantities and can cost up to $100 000 per kilogram. Such enzymes will be required in a high state of purity.

The further growth of world enzyme markets will be polarized around (a) high-volume, industrial grade enzyme products, and (b) low-volume, high-purity enzyme products for analytical, diagnostic or therapeutic applications. In the world production of industrial enzymes it is of interest that two small European countries dominate the markets (Table 5.3).

Among the many new areas of opportunity for enzyme technology is the utilization of lignocellulose (or woody materials) in biotechnological processes. This abundant substrate must be utilized, and many research efforts are being directed to discover new and efficient enzyme systems that can attack the complex molecular configurations of lignocellulose and make available the component molecules. This could well be the most bountiful future area of expansion in enzyme technology.

5.3 The technology of enzyme production

Although many useful enzymes have been derived from plant and animal sources, it is anticipated that most future developments in enzyme technology will rely on enzymes of microbial origin. Even in the malting process of brewing, where the amylases of germinated barley which hydrolyse the starch are relatively inexpensive and around which existing brewing technology has developed, there are now some competitive processes involving microbial enzymes.

The use of microorganisms as a source material for enzyme production has developed for several important reasons.

(1) There is normally a high specific activity per unit dry weight of product.
(2) Seasonal fluctuations of raw materials and possible shortages due to climatic change or political upheavals do not occur.
(3) In microbes a wide spectrum of enzyme characteristics, such as pH range and high temperature resistance, is available for selection.
(4) Industrial genetics has greatly increased the possibilities for optimizing enzyme yield and type through strain selection, mutation, induction and selection of growth conditions and, more recently, by using the innovative powers of gene transfer technology.

Techniques are now available to superproduce enzymes in *Escherichia coli* that are either rare or present only in microorganisms under exacting growth requirements. The ability to transfer these capabilities to more acceptable industrial microorganisms will soon be achieved.

The rationale for selection between different microorganisms is complex, and involves many ill-defined factors such as economics of cultivation, whether the enzyme is secreted in the culture broth or retained in the cell, and the presence of harmful enzymes. Depending on source material, enzymes differ greatly in their stability to temperature and to extremes of pH. Thus *Bacillus subtilis* proteases are relatively heat-stable and active under alkaline conditions and have been most suitable as soap-powder additives. In contrast, fungal amylases, because of their greater sensitivity to heat, have been more useful in the baking industry.

When selecting for enzyme production the industrial geneticist seeks to optimize desired properties (high enzyme yield, stability, independence of inducers, good recovery, etc.), while attempting to remove or suppress undesired properties (harmful accompanying metabolites, odour, colour, etc.). Sophisticated genetic techniques have not yet been widely practised, most manufacturers relying mainly on mutagenization combined with good selection methods. A common feature of most industrial producer organisms is that their genetics are little understood. However, gene transfer technology together with protein engineering may well alter this and present new horizons to enzyme technology.

The raw materials for industrial enzyme fermentations have normally been limited to substances that are readily available in large quantities at low cost, and are nutritionally safe. Some of the most commonly used substrates are starch hydrolysate, molasses, corn steep liquor, whey and many cereals.

Industrial enzyme production from microorganisms relies predominantly on either submerged liquid conditions or solid substrate fermentation as described in Chapter 3.

Solid substrate methods of producing fungal enzymes have long historical applications, particularly in Japan and other Far East countries. In practice, this method uses moist wheat or rice bran with added nutrient salts as substrates. The growing environment is usually rectangular or circular trays held in constant-temperature rooms. Commercial enzymes of importance produced in this way include fungal amylases, proteases, pectinases and cellulases.

Since microbial enzymes are mostly low-volume, medium-cost products, these production methods in submerged liquid systems have generally relied on bioreactors similar in design and function to those used in antibiotic production processes (see Chapter 3). The choice of fermentation medium is important since it supplies the energy needs as well as carbon and nitrogen sources. Raw material costs will be related closely to the value of the final product. A typical enzyme-producing bioreactor is constructed from stainless steel and has a capacity of 10 to 50 m³ (Figs. 5.2, 5.3). In most cases enzymes are produced in batch fermentations lasting from 30 to 150 hours; continuous cultivation processes have found little application in industrial enzyme production. Sterility of the bioreactor system is essential throughout production.

At the completion of the fermentation the enzyme may be present within the microorganism or excreted into the liquid or solid medium. Commercial enzyme preparations for sale will be either in a solid or liquid form, crude or highly purified. The concentration and purification of an enzyme is shown in Fig. 5.4. The final cost of the enzyme will depend on the degree of downstream processing that is required to achieve a saleable product.

All microbial enzyme products used in foods or medically related aspects are required to meet strict specifications with regard to toxicity (Table 5.4). At present only a small number of microorganisms are used for enzyme production. Responsibility for the safety of an enzyme product remains with the manufacturer. In practice, a safe enzyme product should have low allergenic potential, and be free of toxic materials and harmful microorganisms.

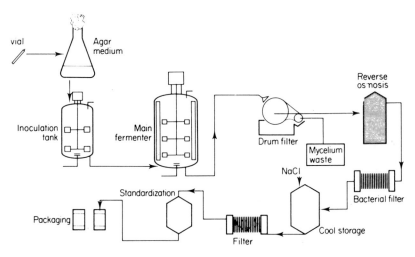

Fig. 5.2 Diagram illustrating the stages in the production of a liquid enzyme preparation.

Fig. 5.3   Fermentation plant at Novo, Denmark, used for the production of enzymes.

5.4   Immobilized enzymes

The use of enzymes in a soluble or free form is wasteful because the enzyme generally cannot be recovered at the end of the reaction. A new and valuable area of enzyme technology is concerned with the immobilization of enzymes on insoluble polymers, such as membranes and particles, which act as supports or carriers for the enzyme activity. The enzymes are physically confined during a continuous catalytic process and may be recovered from the reaction mixture and reused over and over again, thus improving the economy of the process. In essence, this is merely a return to the natural immobilized state of most enzymes in living systems. Some enzymes that are rapidly inactivated by heat when in cell-free form can be stabilized by attachment to inert polymeric supports; while in other examples such insolubilized enzymes can be used in nonaqueous environments. Whole microbial cells can also be immobilized inside polyacrylamide beads and used for a wide range of catalytic functions.

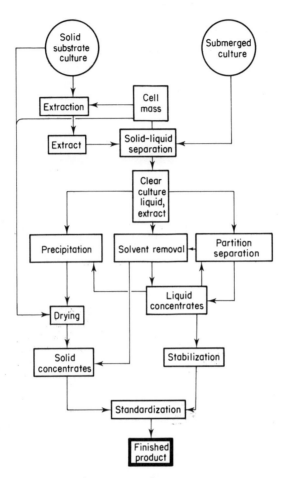

Fig. 5.4 The extraction and preparation of an enzyme.

The variety of new enzymes and whole organism systems that are likely to become cheaply available presents exciting possibilities for the future.

Present applications of immobilized catalysts are mainly confined to industrial processes, for example, production of L-amino acids, organic acids and fructose syrup. The main potential for immobilized biocatalysts lies in novel applications and the development of new products, rather than as an alternative to existing processes using non-immobilized biocatalysts.

Immobilized enzymes are normally more stable than their soluble counterparts, and are able to be reused in the purified, semipurified, or whole-cell form. Catalytic properties of immobilized enzymes can often be altered favourably to allow operation under broader or more rigorous reaction conditions; for example, immobilized glucose isomerase can be used continuously for over 1000 hours at temperatures between 60 and 65°C.

Table 5.4   Safety testing of food enzymes based on the Association of Microbial Food Enzyme Producers classification.

	(a) Microorganisms that have traditionally been used in food, or in food processing	(b) Microorganisms that are accepted as harmless contaminants present in food	(c) Microorganisms that are not included in (a) or (b)
Pathogenicity	In general no testing required		X
Acute oral toxicity, mouse and rat; subacute oral toxicity, rat four weeks		X	X
Three month oral toxicity, rat		X	X
in vitro mutagenicity		X	X
Teratogenicity, rat; *in vivo* mutagenicity, mouse and hamster			(X)*
			(X)*
Toxicity studies on the final food			(X)*
Carcinogenicity, rat; fertility and reproduction			(X)*
			(X)*

X   test to be performed
(X)* only to be performed under exceptional conditions
(from Godfrey and Reichelt, 1983).

How are enzymes immobilized? In practice both physical and chemical methods are routinely used for enzyme immobilization. Physically, enzymes may be absorbed on to an insoluble matrix, entrapped within a gel or encapsulated within a microcapsule or behind a semipermeable membrane. Chemically, enzymes may be convalently attached to solid supports or cross linked.

A large number of chemical reactions have been used for the covalent binding of enzymes by way of their nonessential functional groups to inorganic carriers such as ceramics, glass, iron, zirconium and titanium, to natural polymers such as sepharose and cellulose, and to synthetic polymers such as nylon, polyacrylamide and other vinyl polymers and copolymers possessing reactive chemical

Table 5.5 Limitations of immobilized enzyme techniques.

Method	Advantages	Disadvantages
Covalent attachment	Not affected by pH, ionic strength of the medium or substrate concentration	Active site may be modified; costly process
Covalent crosslinking	Enzyme strongly bound, thus unlikely to be lost	Loss of enzyme activity during preparation; not effective for macromolecular substrates, regeneration of carrier not possible.
Adsorption	Simple with no modification of enzyme; regeneration of carrier possible; cheap technique	Changes in ionic strength may cause desorption; enzyme subject to microbial or proteolytic enzyme attack
Entrapment	No chemical modification of enzyme	Diffusion of substrate and product from the active site; preparation difficult and often results in enzyme inactivation; continuous loss of enzyme due to distribution of pore size; not effective for macromolecular substrates; enzyme not subject to microbial or proteolytic action

(From Atkinson and Mavituna. *Biochemical Engineering and Biotechnology Handbook*, 1983).

groups. In many of these procedures the covalent binding of enzymes to the carriers is nonspecific, i.e. the binding of the enzyme to the carrier by way of the enzyme's chemically active groups is distributed at random. More recent studies have attempted to develop techniques of enzyme immobilization in which the enzyme binds to a carrier with high activity without affecting its catalytic activity. The limitations of immobilized enzyme techniques are shown in Table 5.5.

The entrapment of enzymes in gel matrices is achieved by carrying out the polymerization or precipitation/coagulation reaction in the presence of the enzyme. Polyacrylamides, collagen, silica gel, etc., have all proved to be suitable matrices, but the entrapment process is relatively difficult and results in low enzyme activity.

Immobilized whole microbial cells are increasingly utilized, and tend to eliminate the tedious, time-consuming and expensive enzyme purification steps. Immobilization of whole cells is normally achieved by the same methods as for cell-free enzymes. The greatest potential for immobilized cell systems lies in replacing complex fermentations such as secondary product formation, (i.e.

Table 5.6   The advantages of immobilized biocatalysts.

1   Permits the reuse of the component enzyme(s).
2   Ideal for continuous operation.
3   Product is enzyme-free.
4   Permits more accurate control of catalytic processes.
5   Improves stability of enzymes.
6   Allows development of a multienzyme reaction system.
7   Offers considerable potential in industrial and medical use.
8   Reduces effluent disposal problems.

semisynthetic antibiotics), in the continuous monitoring of chemical processes (via enzyme electrodes), water analysis and waste treatment, continuous malting processes, nitrogen fixation, synthesis of steroids and other valuable medical products. The advantages of using immobilized biocatalysts are summarized in Table 5.6.

As a consequence of successful immobilization techniques in the form of enzyme capsules, enzyme beads, enzyme columns and enzyme membranes many types of bioreactors have been developed at a laboratory scale and to a lesser extent at industrial scale. These include batch-fed stirred tank bioreactors, continuous packed bed bioreactors and continuous fluidized bed bioreactors.

In industrial practice the catalytic properties of isolated enzymes, immobilized enzymes or immobilized whole cells are generally utilized within the confines of bioreactor vessels. Bioreactor systems can have many forms depending on the type of reaction and the stability of the enzyme.

In Europe immobilized penicillin acylase is used to prepare 6-aminopenicillanic acid (6APA) from naturally produced penicillin G or V. This compound is an important intermediate in the synthesis of semisynthetic penicillins, so essential in our fight against bacterial diseases. At least 3 500 tonnes of 6APA are produced each year, requiring the production of about 30 tonnes of the enzyme.

Immobilized glucose isomerase is used in the USA, Japan and Europe for the industrial production of high-fructose syrups by partial isomerization of glucose derived from starch. Millions of tonnes of high-fructose syrup are produced annually using this enzymes, which is undoubtedly the most widely used of all the immobilized enzyme. The industrial and commercial success of this process is due to the following facts: glucose derived from starch is relatively cheap; fructose is sweeter than glucose; the high-fructose syrup contains approximately equivalent amounts of glucose and fructose, and from a nutritional aspect is similar to sucrose.

Another important use of immobilized enzymes is in aminoacylase production of amino acids. Aminoacylase columns are used in Japan to produce hundreds of kilograms of L-methionine, L-phenylalanine, L-tryphophan and L-valine.

Enzyme polymer conjugates are used extensively in analytical and clinical chemistry. Immobilized enzyme columns or tubes can be used repeatedly as specific catalysts in assays of substrates. Enzyme electrodes are designed for the potentiometric or amperometric assay of substrates such as urea, amino acids, glucose, alcohol and lactic acid. The electrode is composed of an electrochemical sensor in close contact with a thin permeable enzyme membrane capable of reacting specifically with the given substrates. The embedded enzymes in the membrane produce oxygen, hydrogen ions, ammonium ions, carbon dioxide or other small molecules depending on the enzymatic reactions occurring, which are readily detected by the specific sensor; the magnitude of the response determines the concentration of the substrate.

The application of enzyme technology to existing processes, for example brewing, food processing, pharmaceuticals, the chemical industry, waste treatment, etc., has enormous potential, but full realization is still a long way off.

Looking to the future, it seems reasonable to expect that the production and application of enzymes will continue to expand. The growing world concern about the environment and resources, in particular the rising prices of oil and other raw materials, is promoting new avenues of research and there is little doubt that enzymes will play a major role in solving these problems.

6

Single Cell Protein Production (SCP)

6.1 The need for protein

A major problem facing the world, in particular the developing nations, is the explosive rate of population growth. At present there are approximately four billion mouths to be fed and with the present rate of population growth this number will exceed five billion by the year 2000. Conventional agriculture may well be unable to supply sufficient food – in particular, protein – to satisfy such demands. The Food and Agriculture Organisation (FAO) already predicts a widening of the protein gap between developed and developing countries. At least 25% of the world's population currently suffer from hunger and malnutrition; a disproportionate number of them live in the developing nations where arid climates and infertile lands hamper productive agriculture.

However, productivity is increasing throughout the world in all branches of agriculture. Biotechnological innovations will accelerate this trend. Food surpluses are occurring in many places, particularly in North America and Western Europe where there are near static populations. Furthermore, some countries that had been net importers of major foods such as India and Indonesia are now self-sufficient. World supply of grain per head has outpaced population growth. However, there are still major imbalances in the availability of cereals, these are further disturbed by changes in global weather patterns (in particular, rainfall), and by national and international warfare with ensuing disruption of agriculture and food distribution.

The extent of the protein problem varies from country to country and must be considered within the framework of each national economy. The shift from grain to meat diets in developed and developing countries is of dramatic proportions and is leading to a much higher *per capita* grain consumption, since it takes 3 to 10 kilograms of grain to produce 1 kilogram of meat by animal rearing and fattening programmes.

The search for sources of protein is relentlessly pursued. New agricultural practices are widespread; high protein cereals have been developed; the cultivation of soybeans and groundnuts is ever-expanding; protein may be extracted from liquid wastes by ultrafiltration; and now the use of microbes as protein producers has gained wide experimental success. This field of study has become known as *single cell protein production* or SCP, referring to the fact that most of the microorganisms used as producers grow as single or filamentous individuals rather than as complex multicellular organisms like plants or animals.

Eating microbes may seem strange, but people have long recognized the nutritional value of some microbes, namely the mushrooms. However, even here, scepticism and prejudice have influenced people's attitudes, and while in some countries mushrooms are widely consumed, in others they are avoided and neglected. Although massive mushroom production processes are now commercially successful throughout the world, this chapter is concerned with the growth of more simple microorganisms, namely bacteria, yeast, fila-mentous fungi and algae, which lend themselves to biotechnological processing. Whereas mushroom growing because of its antiquity can be consi-dered as a conventional type of food production, the use of other microbes is less appreciated and presents many problems, not all of which are technological in nature.

During the last two decades there has been a growing interest in using microbes for food production, in particular for feeding domesticated animals. It has been argued that the use of SCP derived from low-value waste materials for animal feed would improve human nutrition, by taking protein-rich vegetable foods out of the human/animal competition and making them more freely avail-able for human consumption in the producer countries, which are often developing countries. Many major companies throughout the world have long been actively involved in these processes, and many worthwhile products are commercially available.

Protein quality and quantity are the goals of SCP production. However, the microbes also contain carbohydrates, fats, vitamins and minerals, and produce them from (in general), otherwise inedible or low-quality waste material.

SCP may be used as a protein supplement for humans and animals. With humans it has been considered as a protein supplement or as a food additive to improve flavour, fat binding, etc. Because humans have a limited capacity to degrade nucleic acids, additional processing is required before SCP can be used in human foods. In animal feeding it can serve as a replacement for such tradi-tional protein supplements as fishmeal and soymeal. The high protein levels, bland odour and taste of SCP, together with ease of storage, confer considerable potential to SCP in food and food outlets. Its high protein content makes its use attractive in aquaculture, e.g. farming shrimps, prawns, trout, salmon, etc.

Microorganisms produce protein much more efficiently than any farm animal (Table 6.1). The protein-producing capacities of a 250-kg cow and 250 g of microorganisms are often compared. Whereas the cow will put on 200 g of protein per day, the microbes, in theory, could produce 25 tonnes in the same time under ideal growing conditions. However, the cow also has the unique ability to convert grass into protein-rich milk. After decades of research no rival method for that conversion process has been developed. The cow has recently been described as 'a live, self-reproducing, and edible fermenter'.

The advantages of using microbes for SCP production are outlined in Table 6.2.

Table 6.1 The time required to double the mass of various organisms.

Organism	Time
Bacteria and yeasts	20–120 minutes
Moulds and algae	2–6 hours
Grass and some plants	1–2 weeks
Chickens	2–4 weeks
Pigs	4–6 weeks
Cattle (young)	1–2 months
Humans (young)	3–6 months

Table 6.2 The advantages of using microbes for SCP production.

1　Microorganisms can grow at remarkably rapid rates under optimum conditions; some microbes can double their mass every 0.5 to 1 hour.

2　Microorganisms are more easily modified genetically than plants and animals; they are more amenable to large-scale screening programmes to select for higher growth rate, improved acid content, etc., and can be more easily subjected to gene transfer technology.

3　Microorganisms have relatively high protein content and the nutritional value of the protein is good.

4　Microorganisms can be grown in vast numbers in relatively small continuous fermentation processes, using relatively small land area, and are also independent of climate.

5　Microorganisms can grow on a wide range of raw materials, in particular low-value wastes, and can also use plant-derived cellulose.

6.2 Acceptability and toxicology of SCP

A unique aspect of the SCP field is the extent to which it has been influenced by factors other than purely technological or economic. Geographic, political, sociological and psychological influences have shaped the course of development to a very marked degree. In particular, a tremendous amount of attention has been given to the problems of safety, nutritional value and acceptability of the product.

The nature of the raw materials used in SCP processes represents the main safety hazard, for example, the possibility of carcinogenic hydrocarbons in the gas-oil or *n*-paraffins, of heavy metals or other contaminants in the mineral salts, of solvent residues after extraction and of toxin (mycotoxin) production by certain fungi. The process organism must be nonpathogenic and nontoxigenic and the products of the organism's metabolism must be innocuous. Rigorous sanitation and quality control procedures must be maintained throughout the process to avoid spoilage or contamination by pathogenic or toxigenic microorganisms.

Toxicological testing of the final product must include short-term acute toxicity testing with several different laboratory animal species, followed by extensive and detailed long-term studies, two years or more, in both rodent and other species. All in all, it represents a major scientific and financial investment, but after the thalidomide disaster it is better to err on the side of caution.

The acceptability of SCP when presented as a human food depends not only on its safety and nutritional value, but also on other factors. In addition to the general reluctance of people to consume material derived from microbes, the eating of food has many subtle psychological, sociological and religious implications. In various cultures there are many specific associations related to eating, and status and symbolic values are attributed to different kinds of food. More obvious factors influencing acceptability must be considered, such as odour, colour, taste and texture. Thus, if SCP is to be used directly as human food, the skills of the food technologist will be greatly challenged. Most realistic considerations would place SCP for human food at a relatively low priority. Demand will mainly be as a food source for various types of domesticated animals, poultry and fish. However, some industrial processes are primarily directed to human consumption, for example, Rank Hovis McDougall/ICI fungal protein.

What about the political and sociological influences on the production of SCP? These have mainly centred around the use of oil-derived materials as substrate for SCP. Such 'petroproteins' have been suspected of being contaminated with carcinogens, and consequently massive SCP programmes in Japan, Italy and Britain were abandoned and efforts redirected to the study of alcohol/methanol-based SCP and SCP from organic wastes.

The critical parameters of SCP processes are strongly interdependent. The choice of substrate will reflect local political and economic factors and the availability of alternative outlets. The choice of organism will partly determine the process technology and the nature of the product. Acceptability of product will depend in part on the substrate. Because of the large volumes that will be involved in SCP processes, continuous culture techniques will be widely selected for economic reasons. Most large concerns actively involved in SCP production have based their processes on such techniques.

6.3 SCP derived from high-energy sources

Materials with a high commercial value as energy sources or derivatives of such chemicals, for example, gas-oil, methanol, ethanol, methane and n-alkanes, have found wide commercial interest. The microbes involved are mostly bacteria and yeasts, and several processes are now in operation. As would be expected, most oil companies have been or are still involved in this field. The wisdom of using such high-energy compounds for food production has been questioned by many scientists.

Methane as an SCP source has been extensively researched but is now considered to present too many technical difficulties to warrant exploitation.

In contrast, methanol offers great economic interest. A large-scale (75 000-litre) fermentation plant for producing the methanol-utilizing bacterium *Methylophilus methylotrophus* has been constructed by ICI in the UK. Hoechst (West Germany) and Mitsubishi (Japan) are working on a similar process, using yeast strains instead of bacteria. The ICI product (named Pruteen) is being used exclusively for animal feeding. Methanol as a carbon source for SCP has many inherent advantages over *n*-paraffins, methane gas and even carbohydrates composition is independent of seasonal fluctuations; there are no possible sources of toxicity in methanol; methanol dissolves easily in the aqueous phase in all concentrations; and no residue of carbon source remains in the harvested biomass. Several other important technical aspects are also very relevant.

The ICI Pruteen plant is the only process of its kind in the Western world and cannot operate economically at present methanol prices. Methanol represents approximately 50% of the cost of the product. In the US the cost of SCP derived from methanol is 2 to 5 times the cost of fishmeal. In the Middle East the low cost of methanol and higher cost of fishmeal, coupled with a need to produce more animal products, could make SCP an attractive proposition. In the Soviet bloc countries many methanol plants are operated. In part this is due to chronic shortages of animal feeds, excess production of methanol, lack of foreign currency to buy alternatives such as soybean meal and above all a disregard for economic planning. Currently soybean meal retails at about $200 per tonne and fishmeal at $350 to $400 per tonne depending on protein content and quality.

The vast range of studies carried out in the 1960s and 1970s on the potential use of methanol and related compounds as substrates for SCP processes certainly pushed fermentation technology to its limits for cheap bulk product formation. The aerobic process for Pruteen production is the world's largest continuous bioprocess system (see Fig. 3.3 on p. 24). The stringent economies required in these processes led to extensive use of airlift bioreactor design. Furthermore, the massive volume and expense in harvesting and preparing the final product forced many economies of scale and of downstream processing. In the USSR there are now several SCP plants on stream, or shortly to go on stream, with production capacity of between 300 000 and 600 000 tonnes per annum.

Ethanol is a particularly suitable source if the SCP is intended for human consumption. In the foreseeable future the comparative status of ethanol SCP will depend on local factors: overcapacity in ethylene crackers, agricultural carbohydrate surpluses and political decisions about regional economic independence and foreign trade balances.

The use of *n*-alkanes as a substrate for SCP has been extensively studied in many countries and represents a very complex biotechnological process. However, most of these processes have now ceased operation because of suspected health hazards resulting from the presence of carcinogens in the SCP. More recently, however, the massive technology developed in this field in Japan and other Eastern countries has been turned over to the study of alcohol-based SCP and SCP from organic wastes.

6.4 SCP from wastes

The materials that make up wastes should normally be recycled back into the ecosystem, for example, straw, bagasse, citric waste, whey, olive and date waste, molasses, animal manure and sewage. The amount of these wastes can be locally very high and may contribute to a significant level of pollution in water courses. Thus the utilization of such materials in SCP processes serves two functions – reduction in pollution and creation of edible protein.

An attractive feature of carbohydrate waste as a raw material is that if its low cost can be coupled with suitable low process costs, an economic SCP product may be obtained from relatively small operative units. The worldwide trend towards stricter effluent control measures, or parallel increases in effluent disposal charges, leads to the concept of waste as a negative-cost raw material. However, the waste may not be suitable for SCP, or its composition or dilution may be such that transport to a production centre may be prohibitive.

Each waste material must be assessed for its suitability for conversion to SCP. In particular, the level of available technology is important. When a waste is available in large quantities and preferably over a prolonged time, then a suitable method of utilization can be planned (Table 6.3).

SCP processes utilizing waste substrates have been carried out on a commercial scale using various yeast organisms in sophisticated fermenter systems. Substrates and producer organisms include molasses (*Saccharomyces cerevisae*) and cheese whey (*Kluyveromyces fragilis*), while the Symba process developed in Sweden utilizes starchy wastes by combining two yeasts, *Endomycopsis fibuligira* and *Candida utilis*.

The feed value of the yeast produced by the Symba process has been evaluated in vast feeding experiments on different types of animals including pigs, chickens and calves. The animals grew well and no adverse effects were recorded. The Symba process can be conveniently separated into three phases:

Table 6.3 The advantages of using widely available organic wastes for SCP production.

1 Reduces environmental pollution.

2 Most organic wastes are available at low cost in most countries, thus ensuring independence in supply.

3 The wastes are upgraded in energy and protein level.

4 Guards against a protein shortage in a community that may be largely dependent on imports.

5 Allows for technological innovation which can often be transferred to developing countries.

6 Many of the wastes such as cellulose and whey already form accepted parts of animal diet and will avoid the acceptability problems of other unusual wastes, e.g. human wastes and fossil fuels.

Phase 1. Waste water (from, for example, a potato processing plant) containing starch is fed through a heat exchanger and sterilized by steam injection.

Phase 2. Sterilized starch solution is fed through two bioreactors together with the starch-hydrolysing yeast *Endomycopsis fibuligira*. The hydrolysed starch then passes into a large bioreactor with *Candida utilis* as the growing organism.

Phase 3. The harvest stream from the *Candida* bioreactor is passed through vibro-screening and hydrocycloning equipment, then centrifuged. The samples collected can be spray dried and the dried material sifted and bagged or stored in bulk.

Pekilo is a new fungal protein product that is produced by fermentation of carbohydrates derived from spent sulphite liquor, molasses, whey, waste fruits and wood or agricultural hydrolysates. It has a good amino acid composition and is rich in vitamins. Extensive animal feeding test programmes have shown that Pekilo protein is a good protein source in the diet of pigs, calves, broilers, chickens and laying hens. Pekilo protein is produced by a continuous process. The organism, *Paecilomyces variotii*, is a filamentous fungus giving a good fibrous structure to the final product. Fig. 6.1 shows a Pekilo plant.

In Britain, Rank Hovis McDougall, in conjunction with ICI, are now commercially marketing another fungal protein, (mycoprotein) derived from the growth of a *Fusarium* fungus on simple carbohydrates. Unlike almost all other forms of SCP, mycoprotein is produced for human consumption. Again the process is continuous.

Fig. 6.1   The Pekilo plant of Tampella, Finland, producing 10,000 tonnes per annum of fungal protein. Pekilo is a new protein of high nutritional value.

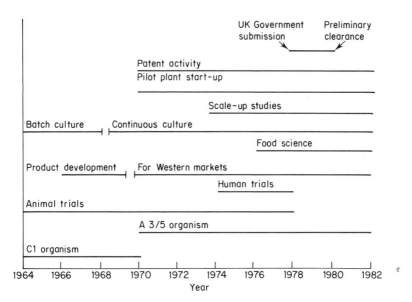

Fig. 6.2 Twenty-year research and development cycle for mycoprotein. Reproduced with permission from Dr. J. Edelman. RIIM Research Ltd.

The research and approval of the Rank Hovis McDougall *Fusarium* myco-protein is estimated to have cost over £40 million and the duration of the project was over 20 years. The original process was batch-fed but then developed into continuous culture. The historical development of the process is shown in Fig. 6.2, and highlights not only the production of the fungal biomass but the equally important human and animal trials, the necessary food science required in formulation and presentation, and the development of market strategies.

A merger between Rank Hovis McDougall and ICI has recently been announced, and it would appear that the advanced technology developed by ICI for the uneconomical Pruteen process may now be used to produce large quantities of the fungal protein. The product has had favourable response from the public. A typical analysis of mycoprotein compared to beef is shown in Table 6.4.

Table 6.4 Typical composition of mycoprotein in comparison to beef.

| Component | % Content (dry weight basis) | |
	Mycoprotein	Raw lean beefsteak
Protein	47	68
Fat	14	30
Dietary fibre	25	Trace
Carbohydrate	10	0
Ash	3	2
RNA	1	Trace

Table 6.5 Strategy for the utilization of complex lignocellulose wastes for SCP production.

1	*Substrate preparation*	(a) Pretreatment of the substrate *Physical:* milling techniques *Chemical:* short acid, alkaline or solvent treatment *Microbial:* screening for rotting organisms (b) Hydrolysis of substrate *Chemical:* complete saccharification with acid or alkali *Enzymatic:* production of cellulases, hemicellulases etc. *Microbial:* screening for good microbial enzyme producers.
2	*Selection and improvement of microorganisms*	
3	*Selection of fermentation systems*	Scale-up, optimization of equipment, process control, batch or continuous process, preservation of final product.
4	*Nutritional and toxicological aspects*	Chemical composition, animal feeding tests, acceptability, toxicology.
5	*Economics*	Cost analysis, comparative considerations with other materials, energy balances.

Cellulose from agriculture and forestry sources and from wastes must constitute the future major feedstock for many biotechnological processes, including SCP. Cellulose, in its natural association with lignin, is by far the most prevalent organic material available for biotechnological conversion. Throughout the world, research teams are studying ways of pretreatment to disrupt or destroy the lignin barrier – physical and chemical methods prevail, but recently a Swedish patent has been taken out for a mutant of *Sporotrichum pulverulentum*, a wood-rotting fungus that selectively utilizes lignin leaving the cellulose relatively undamaged or consumed. In this way the natural cellulose becomes amenable to microbial or enzymatic degradation to simple sugars. The delignified lignocellulose is a worthwhile energy substrate for ruminant animals, which are well able to use cellulose as food. In this way, lignocellulose materials such as straw, bagasse and even wood can become useful animal feedstuffs. The technical approaches to the utilization of complex lignocellulose wastes are outlined in Table 6.5.

Many mushroom-type organisms have long been a source of human food and are produced on lignocellulose materials. These processes are examples of low-energy technological systems. The processes vary not only in the type of substrate and product, but also in the degree of sophistication of methodology. Whereas most SCP processes use liquid fermentation systems, many of the present approaches to lignocellulose degradation rely on a low-water fermentation – known as solid substrate fermentation (see Chapter 3).

Table 6.6   Production of edible fungi by artificial cultivation methods.

Species	Common name	Distribution	Quantity/ kt
Agaricus bisporus	White mushroom	Worldwide	750
Lentinus edodes	Shii-ta-ké	Japan, Far East	180
Volvariella volvaceae	Chinese or straw mushroom	Tropical countries	65
Flammulina velutipes	Winter mushroom	Japan, Taiwan	65
Pleurotus spp.	Oyster mushroom	Worldwide	40
Pholiota nameko	Nameko	Japan	20
Auricularia auricula-judae	'Jew's ear'	Japan, Taiwan	12
Tremella spp. and some other species		Taiwan and worldwide	3
Total			1135

(From Zadrazil & Grabbe, *Biotechnology*, **3**, 45–187 (1983)).

A small portion of the straw produced by agriculture has traditionally been composted with horse manure to produce a substrate suitable for the production of mushrooms (*Agaricus bisporus*). Each year the mushroom industry in the UK consumes about 300 000 tonnes of straw in producing mushroom compost. The technology involves the application of microbiology, fermentation technology and biochemical engineering to the production of microorganisms in a large-scale, agricultural-type process. Mushroom bioprocessing, although recognized as an example of total biotechnology, is still a somewhat neglected area of 'new' biotechnological research. A large number of edible mushrooms are now artificially cultivated throughout the world (Table 6.6).

Biotechnological innovation is only recent in this field but the rewards will be great. Biotechnology need not always be high technology. In developing nations where highly expensive systems may be impractical because of cost and lack of skilled operators, many of the new biotechnological discoveries may well lead to improvements in traditional microbial processes.

Major examples of solid substrate fermentation are the many types of Oriental food fermentations. In many of these processes bland materials such as beans, bran, etc. are subjected to microbial activity, hydrolysing starch and proteins, and creating products with enhanced flavours. Examples are traditional foods such as miso, shoyu, tempeh and many more local formulations. Although many of these microbial-based foods are produced on a 'cottage' scale, others are the basis of large industries demanding major biotechnological inputs. Such foods and flavours are slowly becoming recognized in the West and will surely become a more acceptable part of our daily food intake.

6.5 SCP from agricultural crops

The previous section described how microorganisms can be used to produce SCP from organic waste such as sugars, starch and cellulose. Why not grow certain plants specifically as substrates for SCP processes? The concept of plant-biomass production as a feedstock for biotechnological processes is important, as much higher yields of fixed carbon are attainable utilizing a well-planned, preconceived plantation method than by harvesting natural vegetation or collecting crops or process wastes. At present, such programmes are practised largely for ethanol production, and it is thought that cassava, sugar cane and tapioca palm represent the only crops upon which it is likely that a mainstream fermentation operation could be economically established. When lignocellulose can be economically utilized most parts of the world will have a readily renewable feedstock available for countless processes.

6.6 SCP from algae

There has been some interest in the use of algae as SCP, since they grow well in open ponds and need only CO_2 as a carbon source and sunlight as an energy source for photosynthesis. Algae such as *Chlorella* and *Senedesmus* have long been used as food in Japan, while *Spirulina* is widely used in Africa and Mexico. In some parts of the world algae are used in ponds or lagoons to aid in the removal of organic pollution, and the resultant biomass is harvested, dried and the powder added to animal feed.

6.7 The economic implications of SCP

The economic feasibility of SCP will be dictated by possible uses in competition with comparable existing products. SCP is protein-rich and can be stored and shipped over long distances. Its principal use will be as animal fodder, partly replacing other protein-rich materials such as soybean meal or fish meal. Being biological in the true sense, even though carried out industrially, these processes do not imbalance natural ecosystems. No novel synthetic compounds are produced and the technology, being based on recycling, is pollution-free.

SCP processes are mostly capital and energy intensive, and most processes must be conducted under sterile conditions in expensive equipment that can be cleaned and sterilized. The final product must not be exposed to microbial contamination, particularly by human pathogens. To achieve economies of scale, SCP processes should have an input of at least 50 000 tonnes per year unless operated as a waste-treatment facility in a food processing plant. Thus a considerable volume of raw materials must be close by to meet these production requirements. Water requirements for SCP production are considerable, for both processing and cooling.

The worldwide, large-scale development of SCP processes has contributed greatly to the advancement of present-day biotechnology. Research and development into SCP processes has involved work in the fields of microbiology, biochemistry, genetics, chemical and process engineering, food technology, agriculture, animal nutrition, ecology, toxicology, medicine, veterinary science and economics. In developing SCP processes new technical solutions for other related technologies have been discovered, for example in waste-water treatment, alcohol production and other metabolites, enzyme technology and nutritional sciences.

6.8 Conclusions

There is little doubt that the main short-term impetus for SCP production will come from the increasing legislative requirement for the disposal of both solid and liquid wastes in a manner compatible with the preservation of the environment. The competitiveness of SCP for animal feed will improve when charges for effluent treatment are allowed for. The main reason for uncertainty with regard to profitability is undoubtedly the price of reference proteins, for example soybean and fish meal.

The future of SCP depends heavily on reducing production costs and improving quality. This may be achieved with lower feedstock costs, improved fermentation and downstream processing, and improvement in the producer organisms as a result of conventional applied genetics together with recombinant DNA technologies.

7

Biological Fuel Generation

7.1 Photosynthesis — the ultimate energe resource

The continual depletion of global fossil-fuel energy has generated an ever-increasing need to seek out alternative sources of energy. These have so far included the harnessing of hydro, tidal, wave and wind power, the capture of solar and geothermal energy supplies and the much misunderstood, but most significant, nuclear power. With all of these systems there is yet no definitive answer on both the economic and energetic outlay necessary for successful operation.

There is now a growing appreciation of biological energy systems and that biotechnological advances in this area will soon bring economic reality to selected processes. Biomass such as forest, agricultural and animal residues, industrial and domestic organic wastes, can now be converted by physico-chemical and/or fermentation processes to clean fuels and petrochemical substitutes. As fossil fuel resources are depleted and become increasingly expensive, conversion of organic residues to liquid fuels becomes a more economically attractive consideration. Photosynthetically derived material is not generally in a sufficiently dry state to possess an attractive calorific value, nor in a form that is best suited to modern technology.

Photosynthetic organisms, both terrestrial and marine, can be considered as continuous solar energy converters and are constantly renewable. Plant photosynthesis alone fixes about 2×10^{11} tonnes of carbon with an energy content of 2×10^{21} J, which represents about 10 times the world's annual energy use and 200 times our food energy consumption. The magnitude and role of photosynthesis has gone largely unappreciated because we use such a small proportion of the fixed carbon. Let it not be forgotten that photosynthesis in the past provided all the present fossil carbon sources, namely coal, oil and natural gas. Thus, photosynthetically derived biomass that exists in many available forms in the environment could well be transformed into storable fuels and chemical feedstocks such as alcohols and methane gas. The actual efficiency of solar energy capture by green plants can be as much as 3 to 4%, with the more effective photosynthetic plants like maize, sorghum and especially sugar cane being the most productive.

The current 'energy crisis' that is reverberating throughout the world has focussed attention on the finite nature of fossil fuel reserves. Taken in association with the dramatic increase in industrialization in many Third World countries, this has generated growing economic and trade pressures for cheaper and reliable supplies of energy. The only alternative regenerable supply of feedstocks for the chemical industry will be from the products of photosynthesis, i.e. sugar, starch and lignocellulose.

Biomass can be considered as a renewable energy source, and can be converted into either direct energy or energy-carrier compounds by direct combustion, anaerobic digestion systems, destructive distillation, gasification, chemical hydrolysis and biochemical hydrolysis.

7.2 Sources of biomass

There are three main directions that can be followed to achieve biomass supplies (Fig. 7.1):

(1) Cultivation of so-called energy crops.
(2) Harvesting of natural vegetation.
(3) Utilization of agricultural and other organic wastes.

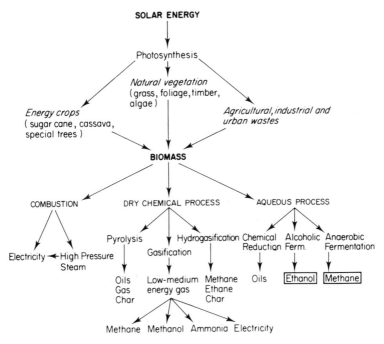

Fig. 7.1 Options for the conversion of biomass to energy.

The conversion of the resulting biomass to useable fuels can be accomplished by biological or chemical means or by a combination of both. The two main end products are methane or ethanol, although other products may arise depending on initial biomass and the processes utilized; for example, solid fuels, hydrogen, low-energy gases, methanol and longer-chain hydrocarbons.

The concept of cultivating plant biomass specifically for energy supply is based on the fact that much higher yields of fixed carbon are attainable from well-planned plantation methods than from harvesting natural vegetation or collecting agricultural or industrial wastes. Programmes of this type are now being extensively planned and practised in many countries throughout the world. Sugar cane and cassava are the two principal crops that are being developed (primarily for ethanol production) in Brazil, Australia and South Africa, whereas more lignocellulosic materials are being developed in Sweden and America. In the latter case plans are being made to grow forests for conversion into liquid fuels. Cost analysis of all of these processes offers considerable encouragement, in particular with sugar cane conversions.

Energy crop plantations will undoubtedly supply meaningful amounts of energy in the near future. The problem of water deficiency is very real, however, and rainfall is most often the limiting factor operating in otherwise ideal conditions of solar radiation intensity, annual hours of sunshine, mild winters and an abundance of good quality land. In certain areas of the world it is possible that such plantations will rapidly become a reality; but for most countries development will centre on the use of organic wastes, namely agricultural, municipal and industrial. Conversion to biofuels could well serve as substitutes for petroleum energy and as a chemical feedstock.

The technical processing of the biomass depends on many factors, including moisture level and chemical complexity. With materials having a high water content, aqueous processing is normal to avoid the need for substrate drying. Alcoholic fermentation to ethanol, anaerobic digestion to methane, as well as chemical reduction to oily hydrocarbons, are all possible. Low moisture level materials such as wood, straw and bagasse can be burnt to give heat or to raise steam for electricity generation; subjected to thermochemical processes such as gasification and pyrolysis to produce energy-rich compounds like gaseous oil, char and eventually methanol and ammonia; or treated by alkaline or biological hydrolysis to produce chemical feedstocks for use in further biological energy conversions.

7.3 Ethanol from biomass

The production of alcohol by fermentation of sugars and starch is an ancient art.

$$C_6H_{12}O_0 \longrightarrow 2CH_3CH_2OH + 2CO_2$$

Production of industrial alcohol by fermentation draws heavily upon the accumulated knowledge of the brewer and the distiller (see Chapter 8). At present industrial alcohol production is largely synthetic, i.e. non-microbial, deriving from petrochemical processes; petrochemical ethanol is made by the hydration of ethylene, and the decline of microbial production of alcohol dates from the large-scale production of ethylene from the 1940s. Within twenty years of development of large-scale petroleum cracking, industrial production of fermentation alcohol fell below potable alcohol production in most industrialized nations. Thus, in technologically more advanced countries ethanol is produced by chemical means. In many developing countries where cheap raw materials are available ethanol is still produced for industrial purposes using traditional fermentation techniques.

A dramatic change in the economics of alcohol production has resulted from the massive increases in the world prices of crude oil. Whereas oil prices have more than quadrupled since 1975, the price of suitable cheap carbohydrates has risen far less on average.

Oil-importing nations are anxious to reduce their import costs and many now subsidize home-produced alternatives. Since alcohol can be used as a partial or complete substitute for motor fuel and can also be converted readily into ethylene and related compounds, its production from indigenous and renewable resources is very attractive.

Nowhere has this been more actively pursued than in Brazil. Vast biotechnological processes operate throughout the country, converting sugar cane and cassava into ethanol by yeast fermentation. A production output of approximately 4×10^6 m^3 ethanol by the early 1980s is now being realized and this will increase to approximately 16×10^6 m^3 by 1990. Brazil's proposed ethanol programme is shown in Fig. 7.2. Brazil's undoubted success in pioneering this production of 'green petrol' is creating worldwide interest, particularly

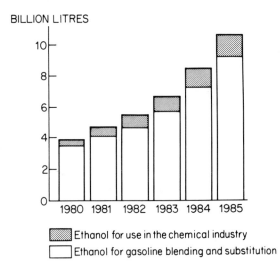

Fig. 7.2 The proposed ethanol programme for Brazil.

amongst poorer Third World nations with the climate and land to grow their own fuel crops but with limited currency to buy oil. Even developed nations such as Australia, USA, Sweden and France, are embarking on biological alcohol production processes utilizing either large agricultural surpluses or forestry wastes. Brazilian officials now estimate that the country's entire petrol needs could be met from planting 0.3% of the country's vast area with alcohol-producing crops. Over 500 fermentation and distillation plants will be built throughout the country processing the crops produced throughout the year. An additional bonus to the energy generation is the creation of over 500 000 new jobs.

The world price of alcohol is around $65 a barrel, still about twice that of oil, and fuel consumption is also up to 20% heavier with alcohol. However, these economic considerations are constantly changing in favour of alcohol production, with ever-increasing oil prices and new design concepts for alcohol-based engines. Furthermore, on a global aspect, green petrol production will help to take some of the pressure off oil products for the rest of the world, reducing competitive tensions and perhaps even wars.

Ethanol for fuel programmes requires a considerable capital investment and will be in keeping with large-scale needs and not small on-farm systems (where methane products are more suitable).

To make available the necessary fermentable sugars, most raw materials require some degree of pretreatment, depending on their chemical composition. With sugar cane this treatment is minimal and consists mainly of the usual milling operation, whereas cassava roots (containing 25 to 38% starch on a wet weight basis) require the action of a suitable saccharifying agent – either acid hydrolysis or enzyme hydrolysis. Cellulosic raw materials such as timber and straw require more extensive pretreatment, and this is reflected in the increased energy inputs required (see Table 7.1). A flow diagram for the production of ethanol from diverse substrates is shown in Fig. 7.3.

Table 7.1 The gross energy requirements of ethanol produced from different substrates by microbial fermentation.

Physical inputs	Sugar cane	Cassava	Substrates Timber[a]	Timber[b]	Straw
Substrate	7.27	19.19	12.67	20.00	4.37
Additional chemicals	0.60	0.89	4.74	6.37	4.74
Water	0.30	0.38	0.80	0.30	0.80
Electricity	7.00	10.47	175.70	7.84	166.74
Fuel oil	8.00	29.03	42.13	62.40	42.13
Capital inputs (buildings etc.)	0.46	1.21	3.34	0.64	3.34
Total	24	61	239	98	222

All figures given in MJ kg^{-1} ethanol.
[a] fermentable sugars formed via enzymic hydrolysis
[b] fermentable sugars formed via acid hydrolysis

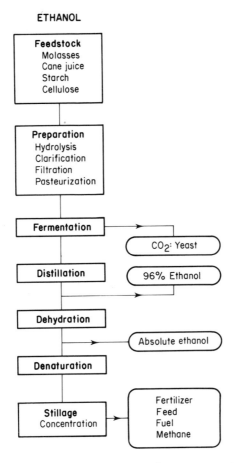

Fig. 7.3 Flow diagram for the production of ethanol (from Coombs, 1980).

The Brazilian programme is almost exclusively based on batch fermentation systems. At present the standards of these fermentations are modest and leave much room for improvement. Continuous methods of production offer many advantages but are really only studied and operated in developed nations with an interest in ethanol formation. Improvements in continuous fermentations have utilized many approaches, including retention of the yeast cells in the bioreactor by separation and recycling and by continuous evaporation of the fermentation broth.

So far, innovation in the Brazilian programme has been restricted to some marginal improvements in essentially traditional alcohol fermentation processes. However, biotechnology is having a considerable input with new developments and numerous research programmes in this field, for example production of more efficient microorganisms by genetic engineering (improved alcohol fermentation, resistance to high temperatures and high alcohol levels, speed of fermentation and higher yields), improved immobilized enzyme reactor technology and by process design improvements. Novel introductions such as fermentation under partial vacuum and recycling of the fermentative yeast cells have increased ethanol productivity to 10 or 12 times that of conventional batch fermentation processes, and such increases reduce capital costs and energy requirements for fermenter operation. Application of these biotechnological improvements to ethanol production will make these processes increasingly economically attractive as a substitute for fossil fuel.

The overall economics of fermenting ethanol from specific crop cultivation in developing countries will achieve support indirectly by the expansion of agriculture, by creating more employment, because oil prices will continue to outstrip agricultural feedstocks, and because new technologies will create further economic uses for the wastes generated in fuel ethanol production.

Vast volumes of wastes or *stillage* result from the Brazilian and other alcohol programmes, and much research is in progress to seek worthwhile end products. Of particular significance will be:

(1) evaporation to feed or fertilizers;
(2) mineralization to ash;
(3) anaerobic fermentation for methanol generation;
(4) conversion by microorganisms into SCP.

7.4 Methane from biomass

The conversion of organic matter to methane by fermentation is a natural process, supplying energy in a clean, high-energy, gaseous state. The natural production of methane occurs in marshes, organic sediments in aquatic systems and in the rumen of cattle.

$$C_6H_{12}O_6 \rightarrow CH_4 + 3CO_2$$

The microbiology of methane production is complex (Fig. 7.4), utilizing mixtures of anaerobic microorganisms. In principle, anaerobic fermentation of complex organic mixtures is believed to proceed through three main biochemical phases, each of which requires specific microbiological parameters. The initial stage requires the solubilization of complex molecules such as cellulose, fats and proteins, which make up most raw organic matter. The resultant soluble, low molecular weight products of this stage are then converted to organic acids; in the final phase of microbial activity, these acids (primarily acetic) are specifically decomposed by the methanogenic bacteria to methane and CO_2.

The most efficient and complex methane-producing system in nature is the rumen. This anaerobic system has never been fully reproduced outside the cow, and is known to be a complex interaction of large numbers of bacteria, protozoa and fungi. All intensively studied bioreactor programmes set up to create methanogenesis under controlled conditions have shown that consistent

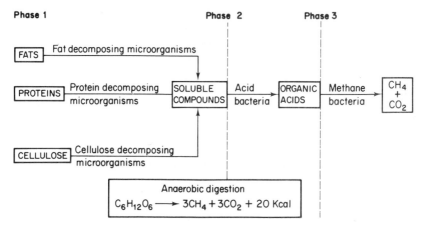

Fig. 7.4   The microbiology of methane generation.

high gas outputs require substantial laboratory monitoring with highly accurate control of environmental variables such as temperature, pH, moisture level, agitation and raw material input and balance. To date, most practical applications of methanogenesis have been at a very low technological level.

There are several possible ways by which methane can be produced in a planned economy: from sewage, from agricultural and urban waste, and in biogas reactors.

The anaerobic digestion of *sewage* is a long-practised technique and many municipal systems have devised methods of capturing the methane and harnessing the energy for the needs of the sewage plant. The energy returns are modest and large-scale expansion does not seem probable.

In recent years methanogenesis of the abundantly available *agricultural and urban wastes* has appeared as an obvious and profitable way to generate energy. The energetic considerations of methane production from such wastes are complex and subject to many limitations. Using urban wastes it should be possible to convert 30 to 50% of the combustible energy to methane, while with the use of certain other vegetable materials or forages it may be possible to achieve 70% conversion. The overall economics of methane production must recognize the valuable by-products generated by the process, namely the effluent and residue rich in ammonia, phosphates and microbial cells which may be used as fertilizer, soil conditioner or even as animal feed. Furthermore, the process can convert malodorous and pathogenic wastes into innocuous and useful materials.

However, there are still many inherent problems that must be overcome before there can be any hope of achieving an energy balance. At present, the cost of collection of organic matter only for the purpose of methanogenesis is too expensive; the rate of methane production is inconsistent and low in most processes, and much research needs to be carried out on the balance of nutrients for process optimization. However, the major problem is the presence of lignin in most agricultural and urban wastes. Lignins are not digested by anaerobic processes, and physical and chemical pretreatment places a considerable energy and cost burden on the overall process.

When methane is produced by the fermentation of animal dung the gaseous products are usually referred to as *Biogas* and the installations called biogas plants or bioreactors. In such systems the animal dung is mixed with water and allowed to ferment in near-anaerobic conditions. Production of biogas by such methods goes back into antiquity and is of particular importance in India, China and Pakistan. The Gobar system of biogas production in the Far East ranges from small peasant systems to quite large plants continuously producing large volumes of gas. In energy terms the simple Gobar system is very near to being a net energy producer on a small scale. Thousands of small-scale plants of family, farm or village size are in operation throughout the world. Larger systems of this type do not achieve a net energy balance.

There has also been some consideration of growing crops on a large scale to provide a 'methane economy'. High-yielding crops in terms of MJ ha^{-1} h^{-1} cultivated on massive land or water areas have been proposed. It has been suggested that 65% of the current gas consumption in the United States could be provided by an energy plantation of area 260 000 km^2 using water hyacinth of energy content 3.8 MJ kg^{-1} dry material. Marine algae have also come under special scrutiny.

Methane generated from organic materials by anaerobic fermentations offers a valuable source of energy that could be directly put to many uses. Furthermore, the associated byproducts may be useful forms of fertilizers for agriculture. Yet, before the full realization of these systems can be achieved, very considerable biotechnological studies must be undertaken. The biological aspects revolve around complex mixed cultures and it is doubtful that the thermodynamic efficiency of the fermentation can be improved. Thus emphasis must be given to improving process design and to technological improvements of the control systems. New and cheaper construction materials for digesters (bioreactors) and gas storage vessels will be required. In time there will be available a complete range of anaerobic technologies to deal with most kinds of biodegradable materials. Although methane will be the principal end product, fuels such as propanol and butanol as well as fertilizers will undoubtedly add to the cost-effectiveness of the overall process.

Methane as an energy source may well have economic value at local small-scale production levels, but there is considerable doubt about the future of large-scale commercial processes for methane production. Some of the more obvious considerations are shown in Table 7.2.

However, anaerobic digestion of municipal, industrial and agricultural wastes can have positive environmental value, since it can combine waste removal and stabilization with net fuel (biogas) formation. The solid or liquid residues can further be used as fertilizer, soil conditioner or animal feed.

7.5 Hydrogen

Consideration has been given to the use of hydrogen as a fuel or in fuel cells for

Table 7.2 Economic arguments against large-scale methane production by microbial processes.

1 An abundance of methane occurs in nature, particularly in natural gas fields and oil field overlays.

2 Methane production by gasification of coal is commercially more attractive.

3 Microbial production of methane is more expensive than natural gas.

4 Costs of storage, transportation and distribution of gaseous fuels is not yet economically worthwhile.

5 Methane cannot be used in automobiles and is difficult and expensive to convert to liquid state.

the production of electricity. Hydrogen production can occur by way of photosynthetic bacteria, biophotolysis of water and by fermentation. In the first two systems, encouraging production of hydrogen has been achieved, but much research is needed to assess the significance of these methods at an applied level. It has been estimated that at least twenty to thirty years of research is needed before any type of functional system is obtained.

Although it is possible to generate H_2 from glucose by bacterial action, the production rate is too small to make microbial genesis of hydrogen economic.

The efficiency of H_2 production by anaerobic fermentation is much less than that of methane production by the same method. Since methane also has a higher energy content it would appear that methane production by microbial processes has a much higher practical potential than hydrogen. However, further research may well alter these considerations.

Although biomass may ultimately only supply a relatively small amount of the world's energy requirements (the estimate for the USA is approximately 5%), it will nevertheless be of immense overall value. In parts of the world such as Brazil and countries of similar climatic conditions, biomass will surely attain wider exploitation and utilization. The technical and agronomic problems are still considerable but biotechnological research is making valuable inroads to further understanding.

Although biomass may still have many disadvantages when compared with oil and coal, the very fact that it is renewable and they are not must be the spur to further research. In time, biomass will become much more easily and economically used as a source of energy.

8

Food and Beverage Biotechnology

8.1 Introduction

The very roots of modern biotechnology are to be found in the fermentation of food and drink, industries which now span almost every society and have evolved over many centuries by empiricism and a strong adherence to tradition (Table 8.1). Alcohol fermentations have ancient ancestry, dating back several thousand years. The present output of the food and beverage industries in both volume and value far exceeds all the new fermentation industries. These industries are usually referred to as traditional biotechnology; microbiological processes have been developed over many years and the resulting products shown to have no obvious adverse effects on humans when consumed in moderation. The safety aspect of the products is of paramount importance and it is

Table 8.1 Food processes involving microorganisms and enzymes.

Traditional Fermentation Processes	Bread
	Beer/wine/alcohol
	Cheese
	Yoghurt
	Meat fermentations
	Vegetable fermentations
Biomass Processes	Food yeasts
	Single Cell Protein (Pruteen, mycoprotein)
Oriental Fermented Foods	Soy sauce
	Tempeh
Glucose and Fructose Syrups	Dextrose, dextrose monohydrates, total sugar
	Fructose syrups
Fermentation for Food Ingredients	Citric acid
	Amino acids
	Flavours and flavour enhancers
	Gums
	Enzymes
	Vinegar

perhaps for this reason that the new areas of biotechnology such as genetic engineering have not been so readily accepted and used.

The food and beverage industries are very different from the pharmaceutical industry; their products are cost and marketing driven rather than technologically driven. Research and development in most of the food and beverage industries is usually less than 1% of sales, is very process-oriented and enjoys little patent protection. Since most food and drink products are high-volume, low-cost items, it is inevitable that market research has become more significant than basic research. Some products such as organic acids, amino acids and gums now increasingly used by the food and drinks industry are in the middle price range, while only a few really high-priced products will have a viable future (e.g. sweeteners and flavourings).

The food and beverage industries are high in terms of turnover and labour employed and are very diverse, ranging from small individual producers to giant multinationals.

The new aspects of modern biotechnology will not revolutionize these industries, but will play an increasingly useful role. Application of new biotechnology will probably generate substitutes for existing preparations rather than entirely new products. Techniques of biotechnology are increasingly contributing to the food and beverage industries' ability to achieve demands for more tailored products and, above all, stringent quality regulations. Techniques of enzyme immobilization and encapsulation and genetic engineering are now beginning to have considerable impact on raw material processing. The potential for development of rapid, inexpensive and highly sensitive biologically based tests for food analyses is considerable. These include enzyme-linked biosensors and monoclonal antibody-based diagnostic kits (see later).

The impact of biotechnology on the food and beverage industries can be anticipated in two directions:

(1) *agronomic*, i.e. increased plant and animal yields, extended growth range and environments from which the farmers will mainly benefit;
(2) *nonagronomic*, i.e. improving plants and microorganisms to provide benefits to the food producer, retailer or consumer (Table 8.2).

New developments in biochemical engineering could also be of advantage to those industries using mechanical (e.g. grinding), physical (e.g. membrane separation, cooking) and chemical (e.g. hydrolysis, salting) methods.

A recent Japanese Industries Association forecast for the year 2000 predicted the value of biotechnologically related markets in the food and drink area as $17 billion, pharmaceuticals $12 billion and commodity chemicals $6 billion. Some of the traditional biotechnologically driven food and beverage industries are briefly out-lined in this chapter, and the impact of new biotechnology on these processes assessed. The food aspects of SCP are discussed in Chapter 6.

Table 8.2   Biotechnology at all levels of the food chain.

Food chain	Potential biotechnological impact
BIOLOGICAL LIVING RAW MATERIALS FOOD RAW MATERIALS FOOD INGREDIENTS	*Agronomic*: increased yield, extend geographic and environmental range, all year growing. *Nonagronomic*: increase benefit to processor by ← lowering the costs of manufacturing operations, keep fresh longer, improve texture and taste, phytoproduction of flavours, colours and other more natural additives, using tissue culture, single cell protein.
FOOD PRODUCTS AT THE FACTORY GATE	By improving processing and reducing product manufacture costs, e.g. starter cultures, enzyme treatments, genetic engineering of microorganisms, detoxification of food 'toxins', upgrading of waste materials, analytical applications and modification of fatty acids, carbohydrates and proteins.
FOOD PRODUCTS AT THE POINT OF CONSUMPTION	Improve distribution and product quality by inhibiting physical, chemical and microbiological deterioration, introducing less harsh processes and new preservation regimes.
PRODUCTS CONSUMED	By ensuring products meet the consumer's expectations of texture, flavour, nutrition, preservation, wholesomeness, and being more natural.

(From Boulter, in *Biotechnology in the Food Industry*, 1–6, 1986)

8.2   Alcoholic beverages

Alcoholic beverages occur throughout the world in many different forms and tastes. The main objective is to produce a controlled quantity of alcohol in the liquid to be harvested after the fermentation. The starting material normally comprises either sugary materials (fruit juices, plant sap, honey) or starchy materials (grains or roots) which need to be hydrolysed to simple sugars before the fermentation (Table 8.3). When these substrates are incubated with suitable microorganisms and allowed to ferment, the end product is a liquid containing anything from a few per cent up to 16% or more of alcohol, with an acid pH and depleted in nutrients for most contaminating microorganisms; these factors combine to give the product a certain degree of biological stability and safety. The alcoholic beverages can be drunk fresh but normal practice requires a period of storage or ageing, leading in many cases to improved organoleptic properties. Further distillation will increase the alcohol strength and

Table 8.3 Substrates for selected alcoholic beverages (nondistilled).

Substrates	Beverage	Country	Saccharifying agent
Starchy (Barley + other cereals)	Ale	Belgium W. Germany Canada Australia	Barley malt
	Lager	Worldwide (industrial countries)	Barley malt
Barley, rye, rice, beet Millet	Kvas Busa Braga Thumba	USSR USSR (Crimea) Roumania India	Barley and rye malt
Rice	Arak Busa Pachwai Saki Sonti	India, SE Asia Turkesta, SSR India Japan India	*Mucor* sp. *Asp. oryzae* *Rhizopus* sp.
Rice (red)	Ancu Hung-Chu	Taiwan China	
Sorghum	Kaffir beer	Malawi	Sorghum malt *Aspergillus* sp. *Mucor rouxii*
	Merissa	Sudan	*Bacillus* spp.
Sweet potato	Awamori	Japan	
Sugary Agave spp. (sap)	Pulque	Mexico	not required
Apple (juice)	Cider	UK, France, N. America	
Grape (juice)	Wine	Temperate; N and S hemispheres	
Honey	Mead	UK	
Pear (juice)	Perry	UK, France	
Palmyra (juice)	'Toddy'	India, SE Asia	
Palm flower-stalk (juice)	Tuwak	Indonesia	

(From Rivière *Industrial Applications of Microbiology*, 1977).

produce spirits of many types, e.g. whisky, brandy, vodka, gin, rum, etc., which contain between 40 and 50% ethanol. Cordials and liqueurs are sweetened alcohol distillates derived from fruits, flowers, leaves, etc.

The most regularly used fermenting microorganism is the yeast *Saccharomyces cerevisiae* or one of its closely related forms. This organism can assimilate and utilize simple sugars such as glucose and fructose and metabolize them to ethanol. It has a high tolerance to ethanol.

$$C_6H_{12}O_6 \rightarrow 2C_2H_5OH + 2CO_2$$

The process details of wine and beer production are briefly examined here since their productions represent major worldwide biotechnological industries.

8.2.1   Wine

Historically, wine is a European drink, and although other parts of the world such as the USA and Australia are now large producers, France, Italy and Germany still produce over half the total world output of approximately 10^{10} litres annually.

Most commercial wines use the wine grape *Vitis vinifera*, and cultivars of this species have been transported throughout the world to establish new wine-producing areas. Soil quality can have an important and subtle effect on the eventual quality of the wine.

Harvesting time of the grapes is judged largely by artisan skills and the grapes are then crushed mechanically or by 'educated' feet. The juice (now termed *must*) is the substrate for the truly biotechnological stage of the production. Since the must will contain many contaminating yeasts and bacteria it is usual practice to add SO_2 to control or abolish this natural fermentation capacity. In large-scale wine production the must is partially or completely sterilized, inoculated with the desired strain of yeast, and subjected to controlled fermentation in suitable tanks or bioreactors. The dryness or sweetness of the wine will depend on the degree of sugar conversion, glycerol levels, secondary infections, etc.

Fermentation conditions such as time and temperature will depend on the type of wine desired. After fermentation, the wines are run into storage vats or tankers where the temperature quickly drops, precipitates form and subtle chemical changes take place. Many wines undergo a spontaneous secondary bacterial (*Leuconostoc* spp.) or malolactic fermentation, converting residual malic acid to lactic acid.

Fortified wines, such as sherry, port and vermouth, are wines to which additional alcohol is added after fermentation, raising the alcohol level to about 20%.

8.2.2   Beer

Beers, ales and lagers are produced mostly from starchy cereals such as barley. Additional carbohydrate sources, known as adjuncts, are normally added in varying proportions. In practice, there are five major steps in the manufacture of beers from grains: malting, mashing, fermentation, maturation and finishing.

Malting. Dried barley is soaked or steeped in water and then spread out on the malthouse floor or in revolving drums, where the seeds germinate with the formation of starch-degrading (amylase) and protein-degrading (protease) enzymes. The germinated seeds are then killed by kilning (slow heating to 80°C) while still retaining most of the enzyme activity (*malt*).

Mashing. In this stage the malt is mixed with hot water, and the starches and proteins break down to produce dextrins, maltose and other sugars, protein breakdown products, minerals and other growth factors (the *wort*). This is the medium for the beer fermentation. Hops may be added prior to the fermentation to give characteristic flavour and some antiseptic properties.

Fermentation. The wort is contained in a bioreactor and inoculated with pure strains of yeast. Yeasts used in brewing are classified as 'top yeasts' or 'bottom yeasts'. Top yeasts float to the surface of the fermenting mixture (ales), while bottom yeasts settle to the bottom of the fermentation tank (lagers).

Maturation and finishing. Beer is usually matured in casks at 0°C for several weeks to improve flavour, settle out the yeasts and remove haze. Bottled or canned beers are usually pasteurized at 60 to 61°C for 20 minutes. The alcohol content of beer is usually not more than 6%; with ales it is somewhat higher.

Traditional applied genetics, together with protoplast fusion and recombinant DNA technology, are constantly improving the yeast strains used in these fermentations. In particular, there has been a vast upsurge in genetic engineering knowledge of *Saccharomyces cerevisiae*.

8.3 Dairy products

The origin of the development of dairy products such as fermented milk, butter and cheeses, is lost in antiquity. It is now known that these fermentations result largely from the activity of a group of bacteria called lactic acid bacteria. Fermentation by lactic acid bacteria results in preservation and transformation of milk and has been used unknowingly for thousands of years. In the past these fermentations arose directly from the natural occurrence of lactic acid bacteria, but gradually it was recognized that a portion of a previously successful 'ferment' when added to milk gave better results. Nowadays, an inoculum (a pure starter culture) of selected bacteria is generally added to the milk to be fermented. The modern worldwide dairy industry owes much to the development of pure starter cultures, good fermentation practices and strict adherence to hygienic protocol.

The lactic acid bacteria can have many beneficial effects in the foods in which they grow.

(1) They have an inhibitory effect on many undesirable bacteria while they themselves are generally harmless; in this way they preserve the milk.
(2) They produce highly acceptable texture and flavour modifications in the milk.
(3) Reputedly, they have beneficial health effects on intestinal microflora.

When growing in milk, these beneficial bacteria break down lactose to lactic acid; however, many other reactions can occur, depending on the composition of the substrate, additives and mode of fermentation. These can result in many other metabolites being formed, giving distinctive flavour and appearance to the milk products, e.g. buttermilk, sour cream, yoghurt and the vast range of cheeses.

One of the largest activities of the dairy industry is cheese production. Cheese is made by separating the casein of milk from the liquid or *whey*. Over 500 kinds of cheese are recognized; yet they could all be prepared from any given batch of milk by proper control of the fermentation and by correct selection of the promoting microorganisms. The production of cheese involves inoculation of milk with a suitable starter culture to generate lactic acid production. A protein-coagulating enzyme, rennet, derived from the stomach of calves, is added to increase coagulation or curdling of the proteins, and the separated curd is cut into blocks, drained and pressed into shapes. Due to shortages of calf enzyme, proteinases or rennet from microfungi (e.g. *Mucor michei*) are increasingly used. A consequence of using microbial instead of calf enzyme is that such cheeses are suitable for vegetarian and kosher diets. Salt may be applied to the curd or to the outside of the pressed cheese.

The flavour of raw cheese is bland and the texture rubbery. It is the period of ripening or maturation, when other microorganisms such as bacteria and fungi can have pronounced effects, that causes the development of distinctive flavours and aromas as well as major textural changes (Table 8.4). World cheese markets now exceed £14 000 million annually.

Traditionally, yoghurt is fermented whole milk; the process uses a mixed culture of *Lactobacillus bulgaricus* and *Streptococcus thermophilus* to convert lactose to lactic acid.

Modern genetic methods, including recombinant DNA technology, are now being widely applied to lactic acid bacteria. Many strains of these bacteria show instability of important product-forming characteristics. It is now recognized that some of the genes controlling these properties reside on plasmids within the cells (see Chapter 4). It has become possible to carry out controlled studies transferring genetic information from one strain to another using rDNA tech-

Table 8.4   Principal groups of cheeses.

Unripened cheeses
        Low fat (cottage cheese)
        High fat (cream cheese)

Ripened cheeses
        Hard cheese (internal ripening)
           Ripened by bacteria (Cheddar, and Swiss cheese)
           Ripened by mould (Roquefort and other blue cheeses)
        Soft cheeses (ripening proceeds from outside)
           Ripened by bacteria (Limburger)
           Ripened by bacteria and moulds (Camembert)

(From Carpenter, *Microbiology*, 1967).

niques. These new techniques will be increasingly used by the food technologists.

8.4 Food enzymes

Enzymes are widely used in the food and beverage industries (see Chapter 5 for details). Such enzymes usually come into the upper price category and are normally used in a relatively crude form. For economic reasons many enzymes need to be in a reusable form, normally achieved by immobilization. The most important and widely used biotechnologically produced enzymes include:

(1) Dairy industry: rennins, lactases, proteases.
(2) Brewing industry: proteases, amylases.
(3) Starch to glucose: alpha-amylase, amyloglucosidase.
(4) High-fructose corn syrup: glucose isomerase.
(5) Texture/colour: proteases, pectinases, amylases.

8.5 Sweeteners

The current preference within the technologically advanced countries for low-calorie foods and drinks has created a big demand for low calorific substitutes to the universal sweetener, sucrose. Saccharin, which has been widely used in this context, is 500 times sweeter than sucrose on a weight basis. There has been extensive research to find new, natural, low calorific sweeteners, and biotechnological methods have been used to develop one of the most important additions to this market, aspartame. Aspartame is 200 times sweeter than sucrose on a weight basis and is the principal sweetener used in the recent upsurge in low-calorie soft drinks. The most expensive component in the synthesis of aspartame is the amino acid phenylalanine, which is now produced on a large scale by several companies using fermentation methods. Extensive toxicological studies were made before this new product was permitted to be used. All new biotechnologically derived food products must undergo a programme of regulatory approval similar to that demanded for pharmaceuticals. Approval for aspartame took 10 years.

8.6 Food wastes

Wastes from the food and drinks industries are becoming an increasing problem, particularly to large production centres, because of increasingly strict legislation on the dumping of high-BOD wastes (see Chapter 11). Much effort is now made to use such organic wastes to generate valuable byproducts, while also achieving active waste removal. There is a large market for biological waste treatment systems.

8.7   Miscellaneous microbial products

These products are derived from microbial fermentations and are used as ingredients for food.

8.7.1   Vinegar

Vinegar is an aqueous solution containing at least 4% acetic acid and small amounts of esters, sugars, alcohol and salts. It is usually derived from wine, malt or apple cider. The fermenting bacteria are normally species of *Acetobacter*.

8.7.2   Citric acid

Citric acid is widely used in the food-related industries in fruit drinks, confectionery, jams, preserved fruits, etc. over 100 000 tonnes of citric acid are manufactured annually by fermentation processes involving the fungus *Aspergillus niger* and molasses as substrate. The fermentation can be as static liquid surface cultures in trays, or in deep-tank, large-scale bioreactors.

8.7.3   Amino acids

Amino acids are widely used in the food and beverage industries as flavouring agents. World production levels are in excess of 200 000 tonnes per year, with Japan commanding a major proportion of the market. Glutamic acid and lysine are two amino acids produced by fermentation processes involving the bacteria *Corynebacterium glutamicum* and *Brevibacterium flavum* respectively.

8.7.4   Gums

Extracellular microbial polysaccharides are produced copiously by many microorganisms and have been used in foods to enhance thickening and to form gels. They can stabilize food structure and improve appearance and palatibility. The bacteria species mainly used are *Pseudomonas* spp. (xanthan gums) and *Leuconostoc mesenteroides* (dextrans). Species of *Acetobacter* can produce cellulose which can form the basis of certain oriental foods.

8.7.5   Flavours/flavour enhancers

The best-known flavour enhancer is monosodium glutamate, now largely made by fermentation using natural or engineered microorganisms. Enzymatic degradation of yeast RNA can produce nucleotide derivatives which are powerful flavour enhancers for meat.

Table 8.5 Per capita annual consumption of fermented foods prepared from *Aspergillus* moulds in Japan (1981).

Food	Per capita/year	Total production/year
1 Shoyu	10.1 litres	1 200 000 kilolitres
2 Miso	4.9 kilograms	572 000 tonnes
3 Saki	12.3 litres	1 445 000 kilolitres
4 Mirin	0.6 litres	260 000 kilolitres
5 Shochu	2.2 litres	30 000 kilolitres
6 Rice vinegar	2.5 litres	305 000 kilolitres
Note:		
Beer	39.5 litres	4 656 000 kilolitres
Whisky and other foreign alcoholic beverages	3.7 litres	445 000 kilolitres
Japanese population 117 850 000 (1 October 1981)		

(From Yokotsuka, in *Microbiology of Fermented Foods*, 1985).

8.8 Oriental fermented foods and drinks

Traditional fermented foods in Japan and related countries are *soy sauce*, derived from a complex series of fermentations of soybeans and other additives, fermented soybean paste or *miso*, fermented rice wine or *saki*, rice vinegar and countless others. The Asiatic countries enjoy a rich range of fermented products with high annual per capita consumption (Table 8.5). Many of these traditional fermented products such as soy sauce are becoming widely accepted in Western cuisine. Over 50 000 tonnes of Oriental fermented foods are consumed annually in the USA.

8.9 Mushroom production

Mushroom cultivation is essentially a solid substrate fermentation process and is one of the few examples of a microbial culture in which the microorganisms cultured can be used directly for human food. Mushroom bioprocessing is truly an example of total biotechnology. A large number of edible mushrooms are now artificially cultivated throughout the world (see Table 6.6. on p. 69). The white mushroom *Agaricus bisporus* is the most widely cultivated. The substrate for most of these cultivations is lignocellulose in the form of straw, wood or wood derivatives. It is the only major biotechnological process that uses lignocellulose commercially.

8.10   Bakery processes

Bread and related bakery products are normally produced from wheat flour, water or milk, salt, fat, sugar and the yeast *Saccharomyces cerevisiae*. When these ingredients are mixed and the sugars are fermented, the dough rises due to the formation of CO_2. This 'leavening' or raising of dough is an ancient process and occurs in one form or another in most parts of the world. The alcohol produced during bread fermentation is always lost during baking. The fermentation achieves three primary objectives: leavening (CO_2 production), flavour development and texture changes in the dough. At the end of the fermentation process the risen dough is baked, giving a final product free of living micro-organisms. Modern applied genetics seeks to improve the quality of the yeast organism, leading to improved activity and better flavour and texture of the product.

9

Biotechnology and Medicine

9.1 Introduction

The impact of pharmaceuticals on human health care is an area where biotechnological innovations are likely to have the earliest commercial realization. The long-standing awareness within the health-related industries of biological and biochemical innovations has led to these industries being heavily involved in biotechnological research, particularly genetic engineering. Furthermore, since health-related products are generally high-value, the financial return warrants extensive research investment. However, the considerable time required to develop a modern pharmaceutical product must not be underestimated, and long periods of toxicological testing are necessary before the national regulatory bodies will grant approval for marketing. The cost of achieving this approval can be many millions of pounds, and the product must have a high sales potential to warrant this investment. Many potentially worthwhile products will not appear on the market because it is not in the financial interest of all producing companies to meet such vast costs of gaining approval. Despite this, a whole new range of pharmaceutical products is being generated using recombinant DNA technologies (Table 9.1).

9.2 Regulatory proteins

These are compounds secreted into the bloodstream by glands in the body, such as the pancreas, pituitary and adrenal to regulate various bodily activities. At present their use in medicine relies on chemical synthesis or extraction from animal tissue; all these processes are normally expensive and subject to supply limitation. Genetic engineering may be able to overcome some of these difficulties, and could also be used to form compounds known to occur in the human body but not so far obtained *in vitro* in usable amounts, e.g. interferons. Within this area there has been massive and continued investment on a worldwide basis by all the pharmaceutical companies. This is high-risk biotechnology, but the rewards for the successful companies are vast. Some of the well-studied compounds in this area include insulin, interferons, human growth hormones, neuroactive peptides and lymphokines.

Table 9.1   Potential pharmaceutical products using recombinant DNA technologies.

Products	Potential use
Biological response modifiers	
Hormones	
Insulin	Treatment of diabetes
Human growth hormone	Treatment of dwarfism and acceleration of wound healing
β-endorphins	Treatment of pain
Secretin	Treatment of digestive problems
Interferons	Treatment of viruses and cancer
Biologicals	
Albumin enzymes	Blood expander
Urokinase	Dissolving blood clots
Clotting factors	Treatment of haemophilia
Antibiotics	Treatment of infectious diseases
Vaccines	Immunization for hepatitis A and B, rabies, influenza, malaria and diphtheria.

9.2.1   Somatostatin

One of the earliest successes in transferring a eukaryotic gene into a bacterial cell has been with somatostatin, a human growth hormone. This hormone is extremely difficult to isolate from animals; half a million sheep brains were extracted to give 0.005 g of pure somatostatin. Using the new gene transfer techniques this same amount of hormone can be produced from 9 litres of a bacterial fermentation. Production is still small but it is hoped that scaling-up will bring greater success.

One child in 5 000 suffers from hypopituitary dwarfism resulting from growth hormone deficiency. Current treatment uses hormones extracted from pituitaries of cadavers, and each treatment can cost up to $10 000. Growth hormone produced using recombinant DNA should shortly have regulatory approval. The annual worldwide market is estimated at $100 million.

9.2.2   Interferons

In 1957 two British researchers discovered substances produced within the body that could act against viruses by making cells resistant to virus attack. Most vertebrate animals can produce these substances, known as interferons, and many animal viruses can induce their *in vitro* synthesis and become sensitive to them. Why, then, have the interferons not become the 'penicillins' of virus infections? Primarily this is because only minute amounts of interferon are produced within cells, and it has proved unbelievably complicated to extract and separate them from cellular proteins.

Human interferons are glycoproteins (proteins with attached sugar molecules) and are believed to play a part in controlling many types of viral infections, including the common cold, as well as having potential in controlling cancer. However, the scarcity of these compounds has consistently hampered efforts to understand the extent of their effectiveness.

There are many different types of interferons characteristic of individual species of animals; mouse interferons will respond to mouse cells but not human cells, and vice versa. Furthermore, different tissues from the same species appear to produce different interferons. Thus interferon for human studies must be derived from human cells, and it has been here that the blockage to production has occurred. Most human interferon production has been carried out in Finland using leucocytes from blood, and the small amounts of interferon produced this way may be used for limited clinical tests throughout the world.

So far, studies have shown that interferons can confer resistance to virus infections, and are involved in the body's natural immune reactions even in the absence of viruses. However, much of the current interest in interferons arises from their ability to inhibit cancer in experimental animals. Interferons present a new approach to cancer therapy because they appear to attack the cancer cells by inhibiting their growth, and that of any viruses involved in the cancer process, and they can also stimulate the body's natural immune defences against the cancer cell. Although the limited clinical studies of these compounds have indicated considerable potential in cancer therapy, the restricted supplies have severely hampered conclusive experimentation; this must await greater availability of interferons.

Two sources of interferon are currently available. The first is from human diploid fibroblasts growing attached to a suitable surface, and the interferon produced is widely considered to be the safest available. The second source is from bacteria in which the gene for human fibroblast interferon has been inserted into a plasmid in such a manner that interferon is synthesized and can be extracted and purified.

Lymphokines are proteins produced by lymphocytes (part of the body's immune system) and are considered to be crucially important to immune reactions. They appear to have the capability of enhancing or restoring the immune system to fight infectious diseases or cancer. Interleukin-2 at present offers the greatest potential and is being extensively researched by the major pharmaceutical companies.

Much of the public image of new biotechnology has centred round the interferon study. Sadly, current research suggests that the newly developed pure interferons do not possess outstanding defensive activity against viral diseases and solid tumours. However, more promising results have been obtained against several types of leukaemia. Much experimentation at the level of treatment procedures has yet to be performed, and it is too early to discount interferons as powerful therapeutic agents.

With each of these important compounds it has been possible to achieve a level of realistic pharmaceutical experimental or drug delivery only because recombinant DNA technology enabled the synthesis of gram quantities of the product.

9.2.3 Insulin

Throughout the world there are millions of people who need regular intakes of insulin to overcome the lethal effects of diabetes. Insulin is extracted from pigs and cattle, and it is now generally believed that some of the unfortunate side effects that occur with regular dosage of insulin could be due to certain additional molecules in the animal insulin. If human insulin was available perhaps these side effects would not occur.

Recombinant DNA research has successfully produced human insulin by bacterial fermentation methods. Regulatory approval has been obtained for an Eli Lilly product called Humulin which is being used throughout the world in increasing amounts. In contrast, Novo, a Danish company, has synthesized a product with the human amino acid sequence from porcine insulin by chemical modification of one amino acid.

9.3 Blood products

Products derived from the fractionation of human blood represent the largest volume of biological pharmaceutical products sold, and is believed to exceed $1 billion on the annual world market. The main plasma commodities are human serum, albumin, gamma globulin and antihaemophilic factor. Japan and North America each utilize 25% of the world's blood products. This industry is characterized by large markets and powerful incentives exist for new biotechnological innovation on a worldwide basis. The great international concern about the transmission of the AIDS virus in blood and blood products has increased awareness of the benefits of producing such blood products by recombinant DNA techniques.

9.4 Vaccines and monoclonal antibodies

During the last decade we have witnessed the unravelling of the bewildering process of immune response in human and animal systems. When a foreign molecule (e.g. a microorganism) enters an animal system a remarkable chain of reactions is set in motion which, if successful, will result in the inactivation and exclusion of the invading molecules. This molecular response can in some cases remain in the animal system for many years, giving complete or partial immunity against that type of molecule. In essence, the foreign molecule is the *antigen*, which can elicit a counteracting response, the *antibody*, from the host system.

In general, antigens are proteins, or proteins combined with other substances such as sugars. Polysaccharides and other complex molecules may also act as antigens. In the disease process, antigens usually reside on the surface of the invading microorganism, and trigger the body's defences against it. In this way antibodies are the essence of immunity against disease.

Antibodies are made by special cells throughout the body and it is now recognized that individual animal species, including humans, can produce unbelievable numbers of different antibodies. The antibody-producing cells recognize the shape of particular determinant groups of the antigen and produce specific antibodies in order to neutralize and eliminate the foreign substance. Thus an animal has sufficient antibodies to combat not only the vast array of microbial invasions that can occur but also an unlimited range of synthetic chemicals. In short, the animal system can bind and inactivate almost any foreign molecule that gets into the living system. However, should a particular antigen challenge not be adequately dealt with, then the invading microorganism can rapidly multiply and create imbalance, illness and perhaps death in the susceptible host.

The ability to stimulate the natural antibodies by vaccines has long been known. Vaccines are preparations of dead microorganisms (or fractions of them), or living attenuated or weakened microorganisms, that can be given to humans or animals to stimulate their immunity to infection. When used on a large scale, vaccines have been a major force in the control of microbial diseases within communities.

Vaccines have been developed against many microbial diseases. However, the success and persistance of the antimicrobial effect varies widely between types of vaccines. Thus vaccines for smallpox and poliomyelitis have almost eliminated these diseases on a world scale, while vaccines against typhoid and cholera are still unsatisfactory. It has not yet been possible to produce any fully effective vaccines against such worldwide diseases as gonorrhea, syphilis, serum hepatitis, malaria and many others. A massive worldwide effort is now in progress to develop a vaccine against the AIDS virus.

Vaccine production is a high-cost, low-volume production system that encompasses many basic principles of biotechnology. Scale-up, in particular, is a constant problem when concerned with virus diseases which need to be produced from animal or human cell cultures. New advances in fermenter technology (see Chapter 3) are rapidly revolutionizing this work and should greatly increase vaccine production in the near future.

New methods of antibody production are now being considered. In current practice antibodies are obtained from immunized animals, but this is usually a tedious and time-consuming operation. At the end of the various extraction and purification stages the antibodies are usually weakly specific, available only in small batches and of relatively low activity. Attempts to culture antibody-secreting cells have been unsuccessful, since such cells neither survive long enough nor produce enough antibodies in culture to become worthwhile sources of antibodies. Furthermore, such systems normally produce mixtures of different antibodies.

The production of vaccines to combat human and animal diseases represents an immense market which has been extensively developed by the pharmaceutical industries. At present vaccine quality and efficacy range from excellent to unsatisfactory.

Table 9.2　US market for major recombinant vaccines.

Vaccine	Immunizations (millions of people)	Price, $ (per immunization)	Values of sales, $ (manufacturers' level)
Initial market for new vaccines (total for first two or three years)			
AIDS	40	20	800 000 000
Herpes	20	20	400 000 000
Total			1 200 000 000
Annual market (once recombinant vaccines are established)			
AIDS	2	15	30 000 000
Herpes	1	15	15 000 000
Hepatitis B	3.5	10	35 000 000
Influenza	3.5	10	35 000 000
Pertussis	3.5	10	35 000 000
Polio	3.5	10	35 000 000
Total			185 000 000

(From Anon (1984) US market for human recombinant DNA vaccines. *Genetic Technology News* **4**(6), 6–7).

In viral-derived diseases, vaccines are being developed by recombinant DNA technology against the influenza virus, polio virus, hepatitis B virus, herpes virus and more recently the AIDS virus. Successful vaccines in these important world diseases could mean massive commercial gains for producing companies. The biggest opportunities exist for diseases such as AIDS or herpes where neither a vaccine nor a cure is yet available.

Extensive studies are also in progress with certain bacterial vaccines, as well as vaccines for parasitic diseases. Malaria still remains the most prevalent infectious disease in the world; this complex and demanding problem could well be overcome in the very near future (Table 9.2).

A significant new development in medically related biotechnology has been the ability to produce monoclonal antibodies (see p. 38). A major advance of this technique is that when antibody-producing cells are immortalized and stabilized the secreted antibodies will always be the same from that particular cell line and can be fully characterized to assess their suitability for different applications. In this way suitable antibodies can be produced, scaled-up (either as ascites tumours in mice or by various forms of fermentation technology), in large quantities allowing much greater standardization for diagnostic applications. Monoclonal antibodies are now finding wide applications in diagnostic techniques requiring highly specific reagents for the detection and measurement of soluble proteins and cell surface markers in blood transfusions, haematology, histology, microbiology and clinical chemistry, as well as in other non-medical areas (Table 9.3).

Monoclonals may also be used in the treatment of tumours and possibly to carry cytotoxic drugs directly to the tumour site.

Table 9.3 Monoclonal antibody markets.

1 Cancer diagnosis and therapy
2 Diagnosis of pregnancy
3 Diagnosis of sexually transmitted diseases
4 Prevention of immune rejection of organ implants
5 Purification of industrial products
6 Detection of trace molecules in food, agriculture and industry

In particular, monoclonals have found ready application in *in vitro* diagnostic products which do not need such rigorous safety testing. Diagnosis can be achieved for many diseases including human veneral diseases, hepatitis B and some bacterial diseases. Monoclonals can also be used in pregnancy testing.

On a commercial scale monoclonal antibodies are being produced in 100-litre airlift fermenters (Celltech), by encapsulation in 100-litre fermenters (Damon Biotech) and in perfusion chambers using lymph from live cattle (Bioresponse).

Commercially, monoclonals have been one of the most rewarding areas of new biotechnology. By 1984 in the US alone there were more than 100 companies producing monoclonals, more than 600 monoclonals were available and more than 60 diagnostic kits had been approved.

9.5 Antibiotics

The discovery in 1929 by Alexander Fleming that a fungus called *Penicillium notatum* could produce a compound selectively able to inactivate a wide range of bacteria, without unduly influencing the host, set in motion scientific studies that profoundly altered the relationship of humans to the controlling influence of bacterial diseases. From these studies emerged the antibiotics penicillin, streptomycin, aureomycin, chloramphenicol and the tetracyclines. Many bacterial diseases have largely been brought under control by the use of antibiotics. Pneumonia, tuberculosis, cholera and leprosy, to mention only a few, no longer dominate society and, at least in the developed parts of the world, have been relegated to minor diseases. Griseofulvin, an antibiotic active against fungi, has brought great relief to those infected with debilitating fungal skin diseases such as ringworm.

Antibiotics are antimicrobial compounds produced by living microorganisms, and are used therapeutically and sometimes prophylactically in the control of infectious diseases. Over 4000 antibiotics have been isolated but only about 50 have achieved wide usage (Table 9.4). The other antibiotic compounds failed to achieve commercial importance for such reasons as toxicity to humans or animals, ineffectiveness or high production costs.

Antibiotics were extensively used in medicine from about 1945 with the arrival of penicillin. New antibiotics soon extended the range of antimicrobial control, and antibiotics are now widely used in human and veterinary medicine and (to a lesser extent) in animal farming, where some antibiotics have been shown to increase the weight of livestock and poultry. Antibiotics can also be

Table 9.4 Some economically important antibiotics.

Antibiotic compound	Producer microorganism	Activity spectrum
Actinomycin D	*Streptomyces sp.*	Antitumour
Asparaginase	*Erwinia sp.*	Antileukaemia
Bacitracin	*Bacillus sp.*	Antibacterial
Bleomycin	*Streptomyces sp.*	Anticancer
Cephalosporin	*Acremonium sp.*	Antibacterial
Chloramphenicol	*Cephalosporium sp.*	Antibacterial
Daunorubicin	*Streptomyces sp.*	Antiprotozoal
Fumagillin	*Aspergillus sp.*	Amoebicidal
Griseofulvin	*Penicillium sp.*	Antifungal
Mitomycin C	*Streptomyces sp.*	Antitumour
Natamycin	*Streptomyces sp.*	Food preservative
Nisin	*Streptococcus sp.*	Food preservative
Penicillin G	*Penicillium sp.*	Antibacterial
Rifamycin	*Nocardia sp.*	Antituberculosis
Streptomycin	*Streptomyces sp.*	Antibacterial

used to a limited extent to control plant diseases and to act as insecticides (see Chapter 10).

Antibiotics that affect a wide range of microorganisms are termed *broad spectrum*, for example chloramphenicol and the tetracyclines, which can control such unrelated organisms as *Rickettsia, Chlamydia* and *Mycoplasma* species. In contrast, streptomycin and penicillin are examples of *narrow-spectrum* antibiotics, being effective against only a few bacterial species. Most antibiotics have been derived from the actinomycetes and the mould fungi.

The production of antibiotics has undoubtedly been the most profitable part of the pharmaceutical industries in the industrialized world. The world market for antibiotics in 1981 was £4500 million for penicillins, £500 million for tetra-cylines, and £450 million for cephalosporins. The present processes are highly efficient and have been achieved with little knowledge about the genetics of the producing organisms. This was, in part, due to the lack of an obvious sexual cycle which limited crossbreeding experiments. However, new techniques such as protoplast fusion and gene transfer technologies are leading to the development of new strains with higher productivity, improved stability and possible new products. These improvements have all resulted in continued decreases in overall costs of production (Fig. 9.1). Modifications in production processes may well follow on from the novel fermenter designs that are gaining wider industrial acceptance.

It is regrettable to note that most studies on antibiotics have been concerned with diseases prevalent in the developed nations. Many diseases of developing countries, including many major tropical diseases, have received little attention from the major pharmaceutical industries. In part this may be due to the high level of technology, including specially trained personnel, that are normally associated with antibiotic research and development. More probably, the

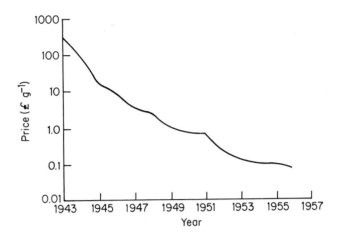

Fig. 9.1 History of the cost of penicillin.

reason lies with the economics of developing new drugs for countries with limited financial resources. Let it be hoped that the advances in biotechnology may make it possible to follow a more enlightened pathway to develop the antibiotics necessary to combat the massive specific disease problems of the developing nations.

A disquieting observation has been the gradual evolution of drug resistance in many bacteria. The possibility of this acquired resistance being transmitted to another species of bacterium is now real. For example, gonorrhea, (a venereal disease) resistant to treatment with penicillin is now present in 19 countries. It is recognized that the resistance factors are located on plasmids within the bacterium and because of this can be more easily transmitted between organisms. The very core of gene transfer technology derives from this phenomenon.

The very large market for antibiotics in animal feeds and food preservation is now under considerable reappraisal. Without doubt, the addition of relatively small amounts of certain antibiotics (for example bacitracin, chlortetracycline, procaine, penicillin, etc.) in the feed of livestock and poultry led to the production of animals that were healthier, grew more rapidly and achieved marketable weight faster. However, there is now little doubt that the incorporation of medically important antibiotics into feed has led to increased spread of drug-resistant microorganisms, increased shedding of dangerous *Salmonella* bacteria in animal dung, and the transfer of antibiotic residues into human food.

As a consequence of the dangers of using antibiotics of human relevance in animal feed, there has been a massive effort to produce antibiotics specifically for animal feed incorporation, so replacing the medically used antibiotics. Thus antibiotics of low therapeutic potency in humans or with an insufficient spectrum of activity are now more regularly used in animal nutrition.

The fight to produce food for the world's ever-increasing population goes on relentlessly. A major part of this strategy involves the battle against microbes.

Biotechnology will play a major part in developing and producing new and effective antibiotic compounds.

The combination of new and traditional technology in the pharmaceutical industry holds outstanding potential for improving microorganisms used in antibiotic production and the isolation of new antibiotic products. The high values of antibiotics and the proportionately low cost of raw materials is a further incentive to research. There is also no competition from chemical processes.

The total worldwide sale of pharmaceutical products was estimated in 1983 at $80 billion of which approximately 10% had major involvement with biotechnology. This was largely made up from antibiotics but with significant contributions from vaccines, steroids, enzymes, vitamins and amino acids. Recombinant DNA technology will significantly increase the proportion of the latter products.

In the past pharmaceutical research has been primarily based on empirical compound testing and clinical observation. New knowledge of biological processes will allow more rational design of new drugs, and recombinant DNA technology will be a major force. This is a highly profitable industry but beset with high-risk products.

10

Biotechnology in Agriculture and Forestry

10.1 Introduction

At present nearly a third of the world's population is hungry, and every day there are over 200 000 more people to feed. Yet only eight countries in the world are able to produce more food than they consume.

The human race is totally dependent on agriculture, and as world populations continue to expand there must be continuous reassessment of agricultural practices to optimize their efficiency. Biotechnology will have its main impact in the agricultural supply industry, providing new strains of plants, improved fertilizers and more efficient and selective pesticides. More antibiotics (not used in human treatments) could be used in animal feeding, improved vaccines could reduce diseases such as foot-and-mouth and swine fever, and growth hormones could be improved. The genetic manipulation and modification of plants, using new genetic engineering principles, is moving forward slowly but confidently and is a priority area of research in most Western nations. There is much research on the use of cloning technology to improve plant resistance to disease and environmental factors such as drought and salinity.

However, agriculture is a politically sensitive area with selective trade barriers and protectionist policies. In many parts of the world, such as the USA and Britain, agriculture is a highly efficient industry and has shown major increases in productivity over recent years. In the long term it is confidently expected that biotechnology will have its greatest commercial impact on agriculture and related areas of forestry and aquaculture. It must always be remembered that agriculture is the largest industry in the world. In the USA, 20% of the total gross national product is contributed by agriculture. The corn seed market alone is $1.5 billion annually.

10.2 Micropropagation and genetic manipulation of plant cells

Biotechnology would appear to be dominated by microbial cell production but it must not be regarded as synonymous with it. Both animal and plant cell culture have a place in the repertoire of the astute biotechnologist.

As early as 1939 it became possible to isolate small numbers of cells from certain plants and to keep them alive indefinitely in artificial cultivation. The cultivation of these tissues required the presence of a plant hormone which allowed the cells to propagate in an unorganized manner resulting in an amorphous mass of cells. Advances were soon made in cultivation techniques, achieving rapid growth rates in chemically defined media. These individuals or groups of cells were treated like microbial suspensions and were able to grow under aerated and shaken conditions, initially in flasks but subsequently in large traditional bioreactors.

The second major advance in plant cell culture was to achieve the complete reversal of this process by causing these individual plant cells to go through a developmental programme from individual cells to tissues, to organs, and finally to entire plants. In this way it has become possible to clone plant cells.

Rapid, large-scale clonal propagation of many plant species including trees is now feasible. Small tissue explants of many species can be aseptically removed from the parent plant, and artifically maintained and increased in number by suitable control of the medium. The process can be rapid and produces high-quality and uniform plants. An outstanding example of this technology has been the recent repetitive cloning of oil palms from callus tissues producing unlimited numbers of stable types. This area of micropropagation not only allows rapid propagation or mass production of identical clones of plant species but also has the following uses (Table 10.1).

(1) Elimination of viruses and other pathogens.
(2) Storage of essential germ plasm instead of conventional seeds.
(3) Embryo rescue.
(4) Production of haploids by anther and ovary culture.

It is now possible to take plant cells and subject them to the battery of manipulative techniques long practised in industrial microbiology, for example mutation, strain selection and process development. Thus the genetic diversity of plants may be altered without the normal sexual process of fertilization, by production of haploid, triploid and tetraploid cells; by the use of protoplast fusion between different species and even genera; and by transformation, i.e. transferring DNA from one plant cell (or even another type of organism) into the cells of another. The techniques of recombinant DNA technology are potentially available to the plant technologist (Fig. 10.1).

Protoplasts can be produced from most plant cells by digesting away the cell wall and maintaining the protoplasts in a suitable osmotic medium. Many types of protoplasts can be induced to reform cell walls and to divide to form cell colonies. Some plants, including potato, pepper, tobacco, tomato, etc., can be fully regenerated from protoplasts. Full regeneration for important cereal crops is not yet possible. Some protoplasts may be fused with other species, allowing a novel mixing of genetic traits.

With tissue culture of callus or large structures, regeneration mainly leads to uniformity of plants. In contrast, regeneration from single plant cells or protoplasts can often be accompanied by minor or extensive changes in the final

Table 10.1 Plant tissue culture technology – current and future applications.

Applied technology	Micropropagation
	Germplasm storage/transport
	Virus elimination
	Embryo rescue
	Haploid production
	Ploidy manipulation
Developing technology	Modification by somaclonal variation
	Gene transfer by protoplast fusion
	Gene introduction by *Agrobacterium*

(From John & Karp, 1985. *Advances in Biotechnological Processes*, 91–121).

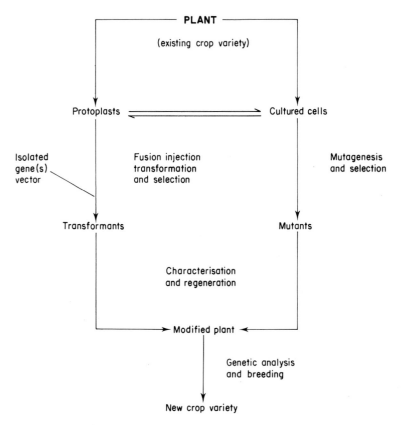

Fig. 10.1 Experimental approach to create a new variety of plant by using biotechnology (OTA, 1984).

plant phenotype. This has been termed *somaclonal variation* and is widely used as a means of crop improvement, in particular with respect to yield and disease resistance.

Genetic engineering could bring the following major improvements to crop production:

(1)   more efficient fertilizer assimilation;
(2)   increased herbicide, microbe and insect resistance;
(3)   enhanced seed oil production;
(4)   increased photosynthetic capacity;
(5)   production of crops better suited to marginal environments, e.g. drought and salinity resistance, etc.

Considerable interest is being shown in using genetic engineering to produce plants with genes to code for toxin production within the plant cells, thus producing plants resistant to invading microbes or insects. This would be cheaper and environmentally safer than present control methods.

The greatest obstacle to the enormous potential of plant genetic engineering is the sheer lack of basic knowledge of the molecular biology of plants. There is a pressing need for more research into plant biochemistry, physiology and molecular genetics.

A soil bacterium, *Agrobacterium tumefaciens*, which can induce tumorous growth in some plants, is now being widely used as a vector for transferring foreign genetic material into plants. It is now possible to remove normal plant genes and to replace them with selected genes. A few years ago people were predicting that regular genetic engineering of plants would not happen before the 1990s. This has not proved so, and many plants can now routinely be altered by genetic engineering. The technology is in place much sooner than expected and will progress rapidly. To date, broad-leaf crops are easier to engineer than cereals.

The ability of seed and plant producers to develop and market new plant types will be a critical feature in plant biotechnology. Because climates vary throughout the world, seed production tends to be a national rather than an international industry. The seed market is one of the largest single markets to which biotechnology is directed. The current world market for seeds is estimated at $30 billion.

It is interesting to compare the future trends of genetic engineering in agriculture as opposed to medicine. In the medical area there are a relatively small number of specific genes that a large number of research groups are studying, whereas in agriculture the number and types of products are virtually limitless. A hugh portfolio of opportunities is available to a wide industrial market. Some important examples have been cited as disease resistance, stress tolerance, herbicide tolerance, changes in product specificity, etc. – could coffee plants be engineered to stop caffeine synthesis and tobacco plants engineered to remove cancer-promoting compounds? In theory all is possible!

Plants are also the source of many important compounds used widely in industry and medicine. Fermenter cultivation of plant cells and concomitant

Table 10.2 Potential markets for plant secondary products.

Compound	Use	Estimated retail market ($ million)
Vinblastine/vincristine	Leukaemia	18–20 (US)
Ajmalicine	Circulatory problems	5–25 (world)
Digitalis	Heart disorders	20–55 (US)
Quinine	Malaria; flavour	5–10 (US)
Codeine	Sedative	50 (US)
Jasmine	Fragrance	0.5 (world)
Pyrethrins	Insecticide	20 (world)
Spearmint	Flavour; fragrance	85–90 (world)

(From Curtin, 1983, Bio technology **1**, 649–672).

product formation could dramatically improve productivity and reliability. Not only may established compounds be produced but possibly quite new compounds may also arise, just as has occurred in the microbial secondary product industries. The range of compounds now produced in static or submerged culture of plant cells is extensive, and future commercial development will relate mainly to the economics of the process. The most likely compounds to find early exploitation will be medicinal, for example opiates, antileukaemic and antitumour agents, and alkaloids and steroids (Table 10.2). Nonmedicinal products will largely be perfumes, flavours and non-nutritive sweeteners. However, the long-term success of these ventures will depend, in part, on whether the genes responsible for such product formation can also be transferred into microbial carriers. If this should happen then it may be more economical to proceed by way of the microbial host cell. Plant cell biotechnology has an exciting contribution to offer to modern society.

10.3 Nitrogen fixation

Although nitrogen makes up approximately four-fifths of the atmosphere, the lack of nitrogen in the soil is the most common limiting factor for plant growth and productivity. Nitrogen must be combined or fixed with other elements such as hydrogen, carbon or oxygen before it can enter and be utilized by plants. Many microorganisms have the ability, either in isolation or in combination, to fix atmospheric nitrogen and in this way contribute immensely to soil fertility. Nitrogen-fixing microorganisms are considered to fix approximately 175 million tonnes of nitrogen each year, equivalent to about 80% of the world's total supply. The remaining 20% is derived from chemically synthesized fertilizers. The global energy crisis has led to vast increases in the cost of fertilizer production; many developing nations can no longer afford to buy and apply nitrogen-based fertilizers, thus aggravating the world food shortage.

Biotechnology will have an important role to play in many of the programmes being developed to increase the role of nitrogen-fixing microorganisms. In particular, genetic manipulation will be used to increase the efficiency of existing systems, and also to transfer the gene complex controlling nitrogen-fixing enzymes into other species, especially agricultural crops such as cereals.

Much publicity has attended a possible transfer by genetic engineering of the genes for nitrogen-fixation into commercial crop plants such as cereals. However, this is a vastly complicated undertaking and worthwhile results must be several decades away. Much fundamental work must yet be done.

Addition of nitrogen-fixing microorganisms such as *Rhizobium* to soil is also strongly advocated in some industrial quarters. However, there is as yet no definitive proof that such additions do give realistic and statistically valid results.

10.4   Biological control

A considerable part of agricultural production is destroyed by insect infestation. As a consequence, numerous toxic, broad-spectrum chemical pesticides have been developed that are highly effective in controlling insects. However, the pollution of the environment by many of these compounds is now widely recognized and several compounds have been withdrawn from commercial use. Alternative methods of controlling insect pests, in particular by the use of microorganisms that cause disease in insects, are now being widely developed and several are already being produced commercially.

Entomopathogens are microorganisms that infect insects, and a wide range of microorganisms including viruses, bacteria, fungi and protozoa have been shown to be potential entomopathogens. The production of these microorganisms in commercial quantities will be a challenging biotechnological endeavour. While many of the entomopathogens, for example bacteria and fungi, can be suitably propagated in conventional surface or submerged fermentation systems, others, for example obligatory parasitic viruses and protozoa, require cultivation in living insects or in cell tissue cultures. The undoubted advances in cell tissue culture methods will most certainly improve the mass production methods for the future.

These new microbial insecticides appear to present little human or environmental hazard, but rigorous testing is a continual prerequisite before release for commercial use (Table 10.3). In practice, they are usually sprayed or dusted on crops. Infected insects may also be released and the environment manipulated to allow the pathogens to exert their maximum effect.

At present, entomopathogens offer considerable potential for controlling the devastating effects of insects in agriculture and forestry. Although many commercial and government agencies are actively involved, this approach to the control of insects is still underexploited.

Some optimistic forecasts predict that biopesticides could capture as much as 40 to 50% of the considerable world pesticide market by the year 2000. A comparison of the efficacy and hazards of chemical insecticides and bioinsecticides is given in Table 10.3. Microbial insecticides are very species-specific and generally will not work unless strict methods of application are used. Although genetic engineering can be used to create nonvirulent strains, they

Table 10.3 Chemical versus microbial pesticides.

	Chemical pesticide	Microbial pesticide
Product use		
Speed of action	Usually rapid	Can be slow
Killing efficacy	Often 100%	Usually 90–95%
Spectrum of effect	Generally broad	Generally narrow
Resistance of target insects	Often develops	None yet shown but microbes are also adaptable
Product safety		
Toxicological testing	Lengthy and costly, c. £3 million	£40 000
Environmental hazards	Many well-known examples	None yet shown
Residues	Interval to harvest usually required	Crop may be harvested immediately.

will require expensive regulatory approval, whereas wild isolates are subject to less stringent registration approval.

Chemical pesticides will remain the stalwarts of pest control for the foreseeable future. However, biotechnological research will allow the experimenter to seek new approaches to the problem of pesticide persistence in the environment and the species spectrum of the pesticide. These will offer entirely new ways to control pests.

Perhaps the skills and novel approaches of biotechnology will lead to a more enlightened commercial use of these microbial insecticides. Continued pressure from those concerned with the safety of the ecosystem should ensure the demise of many of the still widely used and environmentally dangerous synthetic chemical insecticides.

10.5 Agricultural crop production and fermentation

Most biotechnological processes depend, at some stage, upon the availability of carbohydrates. Carbohydrates in their many forms are derived from biomass; and countries with high agricultural and forestry productivity are richly endowed with the basic feedstock for practically every conceivable biotechnological process.

Many agricultural crops are already grown specifically for conversion into other products by fermentation processes, for example grapes for wines, barley for beers, etc. This trend will increase, with particular use being made of sugar cane and cassava for fuel alcohol production, and other substrates for SCP processes. The details of such processes have been discussed elsewhere in this book.

10.6   Biotechnology and forestry

A prerequisite for the full development of biotechnology within the forest industries is a range of microorganisms that can easily attack and breakdown lignocellulose. The main degradation of wood is caused by fungi, and most studies concerned with the selective breakdown of woody materials make use of these filamentous organisms. However, certain bacteria have recently gained prominence in lignin degradation.

Extensive studies concentrate on the use of substances derived from the pulp and paper industries. In particular, two fermentation processes for the production of SCP from waste materials such as sulphite liquor are now well-established commercial processes, i.e. the Symba process and the Pekilo process (see Chapter 6). Not only do these processes produce useable SCP but, more importantly, they aid in reducing the pollution of water courses from the effluent of the pulp and paper industries. Such effluents have been a serious source of environmental pollution and firms must pay enormous costs for waste treatment or sewage system utilization. Biotechnology could offer the pulp and paper industry the opportunity of reduced costs in terms of waste treatment and increased revenue in terms of supplementary product production.

The long-term aim in this field of study is to develop economically sound methods for the complete saccharification of lignocellulose materials for the production of sugar, ethanol and other energy-rich base chemicals. Without doubt, this must represent the most difficult of all biotechnological problems to be solved. Massive worldwide studies are showing good progress and there can be little doubt that the 1990s will witness a biotechnological solution to this problem.

Lignocellulose, the main structural material of trees, may also be chemically modified and used for animal feeding, or used as logs or sawdust for the production of edible mushrooms such as the Japanese Shii-ta-ké mushroom, *Lentinus edodes*.

10.7   Aquaculture

Fish rearing by intensive methods is an important area where the application of the biosciences and process engineering could have significant impact in improving food resources throughout the world. One major problem which is being investigated using biotechnological methods is the control of the diseases that affect the fish in an intensive environment, whether freshwater or marine. Other problems that may have biotechnological solutions are associated with optimal feeding of the fish and the quality of the fish muscle.

Table 10.4 Major biotechnological products being developed for veterinary medicine.

Chicken growth hormone
Bovine growth hormone
Swine growth hormone
Scours vaccines
Foot and mouth vaccines
Newcastle disease vaccine
Rabies vaccine
Monoclonal diagnostics and therapeutics

(From Hacking, 1986).

Table 10.5 Examples of antibiotic compounds of agricultural interest.

Activity	Compound
Anthelmintic	Avermectin
	Hygromycin
Herbicide	Herbicidin
	Anisomycin (NK-049)
Insecticide	Piericidin
	Tetranactin
Miticide	Tetranactin
	Milebemycins
Plant hormone	Gibberellins
Food pigment	Monascin
Detoxicant	Detoxin
Coccidiostat	Monensin
	Lasalocid
Animal growth promotant	Virginiamycin
	Avoparcin
Antiprotozoal	Azalomycin F
Plant disease controller	Blasticidin S
	Validamycin
Food preservative	Nisin
	Natamycin
Piscicidal	Antimycin A
Growth factors	Zearalenol
Antiviral (plants)	Aabomycin A
Abscission agent	Cycloheximide
Fungicide	Cycloheximide

(From Vandamme, 1984, *Biotechnology of Industrial Antibiotics*, 3–32).

10.8 Animal rearing

In animal rearing, biotechnology can be expected to exert considerable influence in several important areas.

Diseases cause vast, almost untold, losses to animal production throughout the world. Recombinant DNA technology is being used to engineer new vaccines to combat disease, and furthermore may be able to do this more cheaply than existing methods (Table 10.4). In this way many hitherto untreatable diseases characteristic of developing countries may be brought under control. Monoclonal antibodies may also be used to diagnose and monitor disease and also to identify the presence of toxic molecules in animal feeds.

Better use of growth hormones (synthesized by biotechnological means) and new and improved feed additives (vitamins, antibiotics) will improve animal health and consequently productivity (Table 10.5). Long-term genetic studies may ultimately lead to strains of animals with improved productivity and greater disease resistance. In both plant and animal agriculture microbially derived antibiotic compounds have a wide beneficial effect.

Agricultural biotechnology is limitless.

11
Environmental Technologies

11.1 Waste-water and sewage treatment

Waste can be considered as any material or energy form that cannot be economically used, recovered or recycled at a given time and place. Growth in human populations has generally been matched by a concomitant formation of a wider range of waste products, many of which cause serious environmental pollution if they are allowed to accumulate in the ecosystem. In rural communities recycling of human, animal and vegetable wastes has been practised for centuries, providing in many cases valuable fertilizers or fuel. However, it was also a source of disease to humans and animals by residual pathogenicity of enteric (intestinal) bacteria. In urban communities, where most of the deleterious wastes accumulate, efficient waste collection and specfic treatment processes have been developed since it is impractical to discharge high volumes of waste into natural land and waters. The introduction of these practices in the last century was one of the main reasons for the spectacular improvement in health and well-being in the developed countries.

Mainly by empirical means a variety of biological treatment systems have been developed, ranging from cesspits, septic tanks and sewage farms to gravel beds, percolating filters and activated sludge processes coupled with anaerobic digestion. The primary aims of all of these systems or bioreactors is to alleviate health hazards and to reduce the amount of oxidizable organic compounds, producing a final effluent or outflow that can be discharged into the natural environment without any adverse effects.

Bioreactors rely on the metabolic versatility of mixed microbial populations for their efficiency. The systems in which they perform their biological functions can be likened to other industrial bioreactors (e.g. antibiotic production); large-scale plants, for example municipal forced aeration tanks (Fig. 11.1), can be extremely complex, requiring the skills of the engineer and the microbiologist for successful operation. The fundamental feature of these biotreators is that they contain a range of microorganisms with the overall metabolic capacity to degrade most organic compounds entering the system.

The development of these systems was an early example of biotechnology. Indeed, in volumetric terms biological treatment of domestic waste-waters and sewerage in the UK is by far the largest biotechnological industry and the least

Fig. 11.1   Aerial view of bioreactors at the sewage treatment plant for the city of Glasgow, Scotland.

recognized by lay people. Controlled use of microorganisms has led to the virtual elimination of such waterborne diseases as typhoid, cholera and dysentery in industrialized communities. Yet, if water and sewage treatments are seriously interrupted, major epidemics may quickly develop as witnessed in 1968 in Zermatt, Switzerland, where typhoid developed following the breakdown of the water treatment plants.

Thus, biotechnology not only generates a whole new range of useful products, it also plays an indispensable part, through water and sewage treatment processes, in the reduction of infectious diseases of humans and animals.

Because of the advances and expansion of modern industrial processes and agricultural practices, the types of waste products have altered radically with time. Many wastes can and do seriously damage the natural environment. In principle, the discharge of industrial wastes should be subject to vigorous government control. The application of restrictive antipollution laws together with the increasing cost of raw materials and energy have created unprecedented new interest in waste treatment and disposal, namely:

(1)   the removal or degradation of toxic materials;
(2)   the production of higher value products such as organic fuels (Chapter 7);
(3)   recycling or recovery of metals, organic materials suitable for fertilizers or soil conditioning, etc.

The biological disposal of organic wastes is achieved in many ways through-

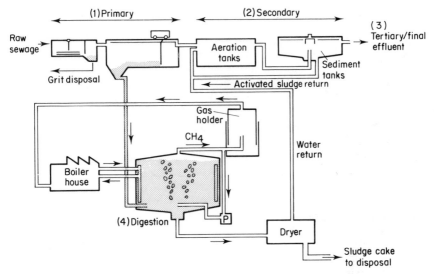

Fig. 11.2   Stages of sewage treatment in a complex incorporating anaerobic digestion.

out the world. A widely used practice for sewage treatment is shown in Fig. 11.2. This complex but highly successful system involves a series of three stages of primary and secondary processing followed by digestion. An optional tertiary stage involving chemical precipitation may be included. The primary activity is to remove coarse particles and solvents leaving the dissolved organic materials to be degraded or oxidized by microorganisms in a highly aerated, open bioreactor. This secondary process requires considerable energy input to drive the mechanical aerators that actively mix the whole system, ensuring regular contact of the microorganisms with the substrates and air. The microorganisms multiply and form a biomass or sludge which can either be removed and dumped, or passed to an anaerobic digester (bioreactor) which will reduce the volume of solids, the odour and the number of pathogenic microorganisms. A further useful feature is the generation of methane or biogas (Chapter 7) which can be used as a fuel. However, the value of biogas is marginal because of its content of carbon dioxide and hydrogen sulphide.

Another important means of degrading dilute organic liquid wastes is the percolating or trickling filter bioreactor. In this system the liquid flows over a series of surfaces, which may be stones, gravel, plastic sheets, etc., on which attached microbes remove organic matter for essential growth. Excessive microbial growth can be a problem, creating blockages and loss of biological activity. Such techniques can also be used in water purification systems.

One of the major contemporary problems has been the release of thousands of new chemicals into the biosphere as products of the synthetic chemical industry. Many such environmentally foreign compounds defy microbial breakdown, and can persist in the ecosystem for indefinite periods, becoming serious environmental hazards – for example, the well-documented persistence of certain pesticides in soils and water courses.

What are the future areas of importance? Microbiological effluent treatment will be a major field of biotechnological interest in the future. Integrated systems will be developed for treating complex wastes. The role of the bio-catalyst or microbe will be constantly reassessed. More effective strains to detoxify dangerous molecules will be developed by conventional genetics and genetic engineering. British scientists have recently developed strains of microbes that detoxify the pesticide Dalopon while other strains can break down 3,5-dichlorobenzoic acids. American scientists have created, through gene transfer technology, a highly active oil-degradative *Pseudomonas* bacterium which offers great prospects for cleaning up oil spillages, etc. The right to patent this organism is now legally accepted and heralds the way for genetic engineers to bring new dimensions to effluent treatment systems.

Systems have now been developed to use immobilized enzymes for the removal of pollutants such as pesticides from water supplies. Whole microbial cell reactors can also be used to concentrate and remove toxic heavy metals.

In countries with high annual hours of sunlight there has been considerable development of combined algal/bacterial systems for waste and water treat-ments. Such processes can lead to the formation of relatively pure water and algal/bacterial biomass, which may be used for animal feeding, biogas forma-tion, or perhaps more ambitiously, for bulk organic chemical formation.

A novel biotechnological innovation in waste-water treatment is the deep-shaft fermentation system developed by ICI. The deepshaft is, in fact, a hole in the ground (up to 150 metres in depth) divided to allow the cycling and mixing of waste water, air and microorganisms (Fig. 11.3). It is most econo-mical in land use and power, and produces much less sludge than conventional systems.

Composting of solid wastes is a long-established practice but has been sadly underdeveloped in the current biotechnology revolution. However, solid substrate fermentation techniques are gaining interest and their potential must await further study.

Finally, the recycling of certain organic effluents, in particular the wastes from agriculture, forestry and the food industry, have been considered in Chapter 6 and elsewhere in the book. In a world becoming increasingly aware of the need to economize on natural resources, energy-rich wastes will find new uses through the technical advances in biotechnology.

11.2   Microbes and the geological environment

Microbes are increasingly recognized as important catalytic agents in certain geological processes, for example mineral formation, mineral degradation, sedi-mentation, weathering and geochemical cycling.

One of the most detrimental examples of microbial involvement with minerals occurs in the production of acid mine waters. This occurs from micro-bial pyrite oxidation when bituminous coal seams are exposed to air and

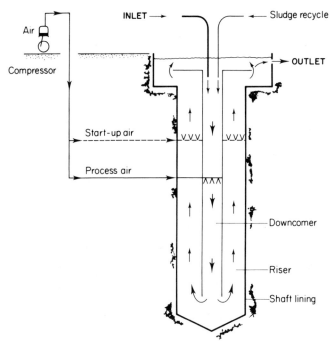

Fig. 11.3   Diagram illustrating the principle of the deep-shaft fermentation system used in wastewater treatment.

moisture during mining. In many mining communities the huge volumes of sulphuric acid produced in this way have created pollution on an unprecedented scale. Other examples of the detrimental effects of microbes include the microbial weathering of building stone such as limestone, leading to defacement or structural changes.

In contrast to these harmful effects, microbes are increasingly used beneficially to extract commercially important elements by solubilization (leaching). For example, metals like cobalt, copper, zinc, lead or uranium can be more easily separated from low-grade ores using microbial agents – mining with microbes.

The biological reactions in extractive metal leaching are usually concerned with the oxidation of mineral sulphides. Many bacteria, fungi, yeasts, algae and even protozoa are able to carry out these specific reactions. Many minerals exist in close association with other substances such as sulphur, iron sulphide, etc. which must be oxidized to free the valuable metal. A widely used bacterium *Thiobacillus ferrooxidans* can oxidize both sulphur and iron, the sulphur in the ore wastes being converted by the bacteria to sulphuric acid. Simultaneously, the oxidation of iron sulphide to iron sulphate is enhanced.

The commercial process or 'fermentation' involves the repeated washing of crushed ore (normally in large heaps, Fig. 11.4) with a leaching solution

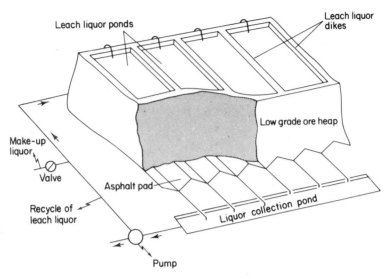

Fig. 11.4   The principle of 'mining with microbes'.

containing live microorganisms and some essential nutrients (phosphate/ammonia) to encourage their growth. The leach liquor collected from the heaps contains the essential metal which can easily be separated (downstream processing) from the sulphuric acid into which it has been extracted.

In the USA almost 10% of total copper production is obtained by this method. Large-scale leaching of uranium ores is widely practised in Canada, India, the USA and USSR. By means of bacterial leaching it is possible to recover uranium from low-grade ore (0.01 to 0.05% U_3O_8) which would be uneconomic by any other known process. The USA alone extracts 4000 tonnes of uranium per year in this manner. Countries such as India, Canada, USA, Chile and Peru are routinely extracting copper at a worldwide annual rate of 300 000 tonnes using microbes.

Continuous processes have been developed, and the control of the essential bacterial populations is easily achieved because of the acidity and limited substrate availability. Leaching technology will continue to offer more efficient and cheaper ways of extracting the increasingly scarce metals necessary for modern industry.

Another important potential application for bacterial leaching is the removal of the sulphur-containing pyrite from high-sulphur coal. Little use is now made of high-sulphur coal because of the SO_2 pollution that occurs with burning. However, as more and more reserves of coal are brought into use, high-sulphur coals cannot be overlooked. Thus the bacterial removal of pyrite (which contains most of the sulphur) from high-sulphur coal could well have huge economic and environmental significance.

Aliphatic hydrocarbon-utilizing bacteria are also being used for prospecting for petroleum deposits. Microbes will soon be commercially used to release petroleum products from oil shelf and tar sands. In all these systems there is rarely any formalized containment vessel or bioreactor. Instead, the natural geological site becomes the bioreactor, allowing water and microorganisms to flow over the ore and to be collected after natural seepage and outflow. Recycling by mechanical pumping can also be used.

Microbes such as the common yeast can also be used to clean up toxic wastes. Radionuclides (e.g. uranium) from waste streams generated by nuclear fuel processing plants have recently been removed by microbial metal adsorption.

In all of these activities multidisciplinary approaches are necessary, and new biotechnological techniques such as designing an organism for a specific function could yield further benefits. The overall picture of this area of biotechnology is one of rapid and exciting development. There is a growing awareness of the value of an unpolluted environment.

12
Safety in Biotechnology

12.1 Introduction

Society has, for centuries, used the products and processes of biotechnology (see Chapter 3). These processes have employed microorganisms of known pathogenic potential and, with the exception of vaccine production, all microorganisms used are nonpathogenic to humans and other animals. Biotechnology has made major advances in public health by the control of communicable diseases with vaccines (Chapter 9) and the improvement in the quality of the environment by the continued improvements of biological waste treatment processes (Chapter 11). Table 12.1 lists the main areas of consideration for safety aspects specific to biotechnology.

12.2 Problems of organism pathogenicity

Many microorganisms can infect humans, animals and plants and cause disease. Successful establishment of disease results from interactions between

Table 12.1 Safety considerations in biotechnology.

---

1 Pathogenicity: potential ability of living organisms and viruses (natural and genetically engineered) to infect humans, animals and plants and to cause disease.

2 Toxicity and allergy associated with microbial production.

3 Other medically relevant effects: increasing the environmental pool of antibiotic-resistant microorganisms.

4 Problems associated with the disposal of spent microbial biomass and the purification of effluents from biotechnological processes.

5 Safety aspects associated with contamination, infection or mutation of process strains.

6 Safety aspects associated with the industrial use of microorganisms containing in vitro recombinant DNA.

(From Kuenzi et al., 1985, Applied Microbiology and Biotechnology **21**, 1-6).

the host and the causal organism. Many factors are involved, only a few of which are well understood.

Most microorganisms used by industry are harmless and many are indeed used directly for the production of human or animal foods. Many such examples have been discussed elsewhere in this book and include yeasts, filamentous fungi and many bacteria. Their safety is well documented from long associations lasting up to hundreds of years. Only a small number of potentially dangerous microorganisms have been used by industry in the manufacture of vaccines or diagnostic reagents, e.g. *Bordetella pertussis* (whooping cough), *Mycobacterium tuberculosis* (tuberculosis) and the virus of foot-and-mouth disease. Stringent containment practices have been the norm.

In recent years there have been many scientific advances permitting alterations to the genetic make-up of microorganisms. Recombinant DNA techniques have been the most successful but have also been the cause of much concern to the public. However, this natural anxiety has been ameliorated by several compelling lines of evidence.

(1) Risk assessment studies have failed to demonstrate that host cells can acquire novel hazardous properties from DNA donor cells (Table 12.2).
(2) More rigorous evaluation of existing information concerning basic immunology, pathogenicity and infectious disease processes has led to relaxation of containment specifications previously set down.
(3) Considerable experimentation has shown no observable hazard.

However, care must always be adopted when using recombinant DNA molecules.

A classification of the degree of potential hazard of microorganisms has been drawn up by the European Federation of Biotechnology (Table 12.3). Group E contains those microorganisms that present risks only to the environment, particularly to animals and plants.

12.3 Problems of biologically active biotechnology products

Vaccines and antibiotics are obvious examples of biologically active products, and care must be taken to prevent their indiscriminate dispersal.

Table 12.2 Risk assessment.

1 Elucidate the capacity of the microorganism to have an adverse effect on humans or the environment.
2 Establish the probability that microorganisms might escape, either accidentally or inadvertently, from the production process system.
3 Evaluate the safety of the desired products and the methods for handling byproducts.

(From Health Impact of Biotechnology, Swiss Biotechnology 1984, **2**, 7–3).

Table 12.3   Proposed European Federation of Biotechnology classification of micro-organisms according to pathogenicity.

Class 1
Microorganisms that have never been identified as causative agents of disease in man and that offer no threat to the environment.

Class 2
Microorganisms that may cause human disease and which might therefore offer a hazard to laboratory workers. They are unlikely to spread in the environment. Prophylactics are available and treatment is effective.

Class 3
Microorganisms that offer a severe threat to the health of laboratory workers but a comparatively small risk to the population at large. Prophylactics are available and treatment is effective.

Class 4
Microorganisms that cause severe illness in humans and offer a serious hazard to laboratory workers and to people at large. In general effective prophylactics are not available and no effective treatment is known.

Group E (Environmental risk)
This group contains microorganisms that offer a more severe threat to the environment than to people. They may be responsible for heavy economic losses. National and international lists and regulations concerning these microorganisms are already in existence in contexts other than biotechnology (e.g. for phytosanitary purposes).

(From Kuenzi *et al.*, 1985, Applied Microbiology and Biotechnology **21**, 1–6).

Contaminants in otherwise safe processes may produce toxic molecules that could become incorporated into final products leading to food poisoning. Allergenic reactions to product formulations must also be guarded against.

Overuse of antibiotics in agriculture could lead to carry-over into human foods, resulting in possible development of antibiotic resistance in human disease organisms. Many countries now restrict the use of antibodies in agriculture.

When properly practised, biotechnology is safe and the benefits deriving from biotechnological innovations will surely lead to major improvements in the health and well-being of the world's population. However, biotechnology must always be subjected to sound regulations for its successful application. The potential risks of biotechnology are manageable, and regulations are being constructed for that management.

13
Conclusions

Biotechnology is a range of biological, chemical and engineering disciplines with varying degrees of application to industrial situations. Biotechnology should be viewed as a spectrum of enabling technologies which find a variety of applications in many industrial and environmental sectors. The assessment of the viability of biotechnology in any country must be made in the context of its biological sciences and their relationships with the productive sectors. Modern biotechnology springs from universities and other research centres that generate the basic knowledge needed to solve practical problems posed by society.

Historically, the applied use of biological organisms (in particular micro-organisms) has developed in a somewhat empirical manner over many years. In many ways the control of the processes was seen more as an art than as a science. In more recent times most of these ancient biotechnological processes have been subjected to rigorous scientific scrutiny and analysis which is, in many cases, influencing and replacing traditional empiricism; of particular significance has been the growth of the science of genetics.

In many long-established biotechnological processes such as brewing, cheese production, etc., organisms that improve the process have been deliberately selected by man. Better understanding of genetics led to more effective application of the genomic potential of the organisms used in these industries. However, in the past ten years, major new developments in the ability to select and manipulate genetic material (recombinant DNA technologies, cell fusions, etc.) have led to unprecedented interest in the industrial uses of living organisms. It is confidently anticipated that these new genetic techniques, coupled with advances in fermentation technology and downstream processing, will have major economic impact on biologically based industries, agriculture and forestry throughout the world. However, although much has been prophesied, only a very small amount of industrial success has been realized to date. Perhaps there has been too much popular emphasis on the highly futuristic aspects of biotechnology; they will almost all be realized eventually, but on a considerably longer time-scale than previously anticipated. In practice, the majority of products to be derived from biotechnology will be recognizable extensions of existing ones or stem from improvements in production. Minor improvements in processes by many of these new technologies may not be highly newsworthy but will help to give the producer companies a competitive edge. The history of biotechnology shows how intimate must be the interplay between industry and academia to maintain the competitive edge to generate new products and to expand its scope, profitability and social impact.

Although there is now a vast reservoir of relevant biological and engineering knowledge and expertise waiting to be put into productive biotechnological use, the eventual rate of application will be determined not primarily by science and technology, but rather by many other equally important factors such as industrial investment policies, the establishment of market needs, economics of new processes, safety factors and government regulations and, above all, the marketing skills needed to introduce new products into commercial use.

Notwithstanding the obvious difficulties, there is no doubt that biotechnology in its many forms will have an increasingly important role to play in our society. However, biotechnology will require highly trained and skilled professionals because its raw materials are knowledge and intelligence.

Glossary

Aerobic   Living or acting only in the presence of oxygen.

Amino acids   The building blocks of proteins.

Anaerobe   Microorganisms that can grow and multiply in the absence of oxygen.

Anaerobic digestion   A microbial fermentation of organic matter to methane and CO_2 which occurs in near absence of air.

Antibiotic   A specific type of chemical substance that is used to fight microbial infections usually in humans or animals. Many antibiotics are produced by microorganisms, some semisynthetically.

Antibody   A protein produced by humans or higher animals in response to exposure to a specific antigen, characterized by specific reactivity with its complementary antigen.

Antigen   A substance, usually a protein or carbohydrate, which when introduced into the body of a human or higher animal, stimulates the production of an antibody that will react specifically with it.

Ascites   Liquid accumulation in the peritoneal cavity, widely used as a method for propagating hybridoma cells for monoclonal antibody formation.

Bacteriophage   A virus that multiplies in bacteria.

Biological oxygen demand (BOD)   The oxygen used in meeting the metabolic needs of aerobic organisms in water containing organic compounds.

Bioreactor (fermenter)   Containment system for fermentation processes.

Biosensor   An electronic device that uses biological molecules to detect specific compounds.

Callus   Plant cells capable of repeated cell division and growth.

Cell line   Cells that acquire the ability to multiply indefinitely *in vitro*.

Conjugation   The transfer of genetic material from one cell to another by cell-to-cell contact.

Continuous fermentation   A fermentation process that can run for long periods, in which raw materials are supplied and products removed continuously.

Chromosomes   The threads of DNA in the nucleus which carry genetic inheritance.

Clone   A group of genetically identified cells descended from one parent.

Downstream processing   Separation and purification of product(s) from a fermentation process.

Enzyme   A class of proteins that control biological reactions.

Fermentation   The process by which microorganisms turn raw materials such as glucose into products such as alcohol.

Gene   A unit of heredity; a segment of DNA coding for a specific protein.

Gene transfer   The use of genetic or physical manipulations to introduce foreign genes into host cells to achieve desired characteristics in progeny.

Genome   The genetic endowment of an organism or individual.

Hybridoma   A unique fused cell which produces quantities of a specific antibody, and reproduces endlessly.

Immobilization   Conversion of enzymes or cells from the free mobile state to the immobile state.

Ligase enzyme   Enzyme used by genetic engineers to join cut ends of DNA strands.

Lignocellulose   The composition of woody biomass, including lignin and cellulose.

Metabolite   Product of biochemical activity.

Mutation   Stable changes of gene inherited on reproduction.

Monoclonal antibodies   Homogeneous antibodies derived from a single clone of cells, a hybridoma.

Plasmid   DNA molecules that can be stably inherited without being linked to a chromosome.

Proteins   Large molecules consisting of amino acids, and the products of genes.

Protoplast   Microbial or plant cell whose wall has been removed so that the cell assumes a spherical shape.

Recombinant DNA   The hybrid DNA produced by joining pieces of DNA from different organisms.

Restriction enzymes   Enzymes used by genetic engineers to cut through DNA at specific points.

Single cell protein   Cells or protein extracts of microorganisms grown in large quantities for use as human or animal protein supplements.

Somaclonal variation   Genetic variation produced from the culture of plant cells from a pure breeding strain.

Splicing   Gene splicing, manipulation, the object of which is to attach one DNA molecule to another.

Scale-up   Expansion of laboratory experiments to full-sized industrial processes.

Tissue culture   A process where individual cells, or clumps of plant or animal tissue, are grown artificially.

Transduction   The transfer of bacterial genes from one bacterium to another by a virus (bacteriophage).

Transformation   The acquisition of new genetic markers by the incorporation of added DNA.

Vectors   Vehicles for transferring DNA from one cell to another.

Further reading

Anon (1984). *Genetic Engineering of Plants*. National Academy Press, Washington.
Broda, P. (1979). *Plasmids*. W.H. Freeman & Co., Oxford.
Bull, A.T., Holt, G. and Lilly, M.D. (1982). *Biotechnology: International Trends and Perspectives*. OECD, Paris.
Cooney, C.L. (1983). Bioreactors: design and operation. *Science*, **219**, 728-33.
Demain, A.L. (1981). Industrial microbiology. *Science*, **214**, 987-95.
Demain, A.L. (1983). New applications of microbial products. *Science*, **219**, 709-14.
Dunnil, P. and Rudd, M. (1984). *Biotechnology and British Industry* Report for SERC, Swindon.
Fishlock, D. (1982). *The Business of Biotechnology*. Financial Times Business Information, London.
Forster, C.F. (1985). *Biotechnology and Waste-water Treatment*. Cambridge Studies in Biotechnology 2. Cambridge University Press, Cambridge.
Hacking, A.J. (1986). *Economic Aspects of Biotechnology*. Cambridge Studies in Biotechnology 3. Cambridge University Press, Cambridge.
Hammond, S.M. and Lambert, P.A. (1978). *Antibiotics and Antimicrobial Action*. Studies in Biology no. 90. Edward Arnold, London.
Higgins, I.J., Best, D.J. and Jones, J. (1985). *Biotechnology, Principles and Applications*. Blackwell Scientific Publications Oxford.
Industrial Microbiology and the Advent of Genetic Engineering (1981). *A Scientific American Book*. W.H. Freeman & Co., New York.
Jacobsson, S., Jamison, A. and Rothman, H. (1986). *The Biotechnology Challenge*. Cambridge University Press, Cambridge.
King, P.P. (1982). Biotechnology: an industrial view. *Journal of Chemical Technology and Biotechnology*, **32**, 2-8.
Old, R.W. and Primrose, S.B. (1980). *Principles of Gene Manipulations - An Introduction to Genetic Engineering*. Blackwell Scientific Publications, Oxford.
OTA (1981). *Impacts of Applied Genetics: Microorganisms, Plants and Animals*. Congress of the United States Office of Technology Assessment, Washington.
OTA (1984). *Commercial Biotechnology: An International Analysis*. Congress of the United States Office of Technology Assessment, Washington.
Postgate, J. (1975). *Microbes and Man*, Penguin, London.
Slesser, M. and Lewis, C. (1979). *Biological Energy Reserves*. E & F.M. Spon, London.
Smith, J.E. (1985). *Biotechnology Principles*. Aspects of Microbiology 11. Van Nostrand Reinhold, Holland.
Stanbury, P.F. and Whitaker, A. (1984). *Principles of Fermentation Technology*. Pergamon Press, Oxford.

Watson, J.S. and Tooze, J. (1981). *The DNA Story*. W.H. Freeman & Co., New York.
Wiseman, A. (ed.) (1983). *Principles of Biotechnology*. Sussex University Press/Chapman & Hall, New York.
Wood, B.J.B. (ed.) (1985). *The Microbiology of Fermented Foods*, vols 1 and 2. Elsevier Applied Science Publishers, London.
Yanchinski, S. (1985). *Setting Genes to Work: the Industrial Era of Biotechnology*. Viking, New York.

New developments in biotechnology are covered in many journals; in particular, reference should be made regularly to *Bio/Technology* (ISSN-22X), Nature Publishing Co., New York.

Index